Lecture Notes in Mathematics

Edited by A. Dold and B. Eckmann

Subseries: Instituto de Matemática Pura e Aplicada, Rio de Janeiro

Adviser: C. Camacho

1259

Felipe Cano Torres

Desingularization Strategies for Three-Dimensional Vector Fields

Springer-Verlag

Berlin Heidelberg New York London Paris Tokyo

Author

Felipe Cano Torres
Departamento de Algebra y Geometría
Facultad de Ciencias
Valladolid 47005, Spain

This volume is being published in a parallel edition by the Instituto de Matemática Pura e Aplicada, Rio de Janeiro as volume 43 of the series "Monografías de Matemática".

Mathematics Subject Classification (1980): 24B05, 32B30, 58A30, 58F14

ISBN 3-540-17944-5 Springer-Verlag Berlin Heidelberg New York
ISBN 0-387-17944-5 Springer-Verlag New York Berlin Heidelberg

Printed in Germany

Printing and binding: Druckhaus Beltz, Hemsbach/Bergstr.
2146/3140-543210

To Mercedes

INTRODUCTION

Let $D = a\partial/\partial x + b\partial/\partial y$ be a plane vector field (or a derivation of a two-dimensional ring of power series over a field k). We can measure how "bad" is the singularity of D at the origin by means of the number $\nu(D) = \min(\nu(a),\nu(b))$, where $\nu(a)$, resp. $\nu(b)$, are the orders at the origin of a, resp. b. Assume moreover that a and b have no common factor (this allows us to consider the "saturated foliation", see |6|). In this situation, after making a finite number of quadratic blowing-ups of the ambient space, we obtain $\nu(D)\leq 1$ at every singular point (see,e.g., Seidenberg's result |13|).

The behaviour of the measure $\nu(D)$ after applying a quadratic blowing-up is not as good as one could expect. For instance, if $D = y\partial/\partial x + x^3\partial/\partial y$ we have $\nu(D)=1$, but the strict transform in a suitable chart is given by $D'= y'x'\partial/\partial x'+(x'^2-y'^2)\partial/\partial y'$ and the order $\nu(D')$ at the origin has been increased by a unit (in general, under any sequence of quadratic blowing-ups the order remains $\leq \nu(D)+1$). This difficulty may be avoided by considering the vector field as a vector field which is " tangent to the exceptional divisor $x'=0$ ". In this case a basis of the free module of such vector fields is given by $x'\partial/\partial x'$ and $\partial/\partial y'$, thus the corresponding coefficients for D' are y' and $x'^2-y'^2$. Now, the adapted order (the minimum of the orders of these coefficients) is one and it has not been increased. If we take this approach, a similar result can be proved: after a finite number of quadratic blowing-ups we obtain that the adapted order is ≤ 1 (see Giraud |8|, or |3| for a different proof). Giraud |9| has shown that this logarithmic point of view is also useful for the problem of reduction of singularities of varieties in positive characteristic.

One can say something more in the above situation: by adding, if necessary, a component to the exceptional divisor, we can always obtain that the adapted order is zero (hence some coefficient is a unit). But this component may be formal and so it can not be globalized (see |3| or I.4.1.4.).

If we consider the problem of the reduction of an n-dimensional vector field $D = \sum a_i\partial/\partial x_i$ from the logarithmic point of view as above, the same obstruction as for n=2 indicates that the statement for the "global reduction result" would be to reach adapted order ≤ 1. Moreover, in higher dimension we can not expect to obtain the reduction of the adapted order only by means of quadratic blowing-ups: thus we have to introduce the permissible centers which will be used in the process. These centers are defined from a local point of view and in this way two problems arise (like in the reduction of singularities of varieties): the existence of an algorithm defined locally in each step and the globalization of this algorithm. These notes are devoted to the first problem in the case $n = 3$.

Technically, the existence of such an algorithm will be formulated as the existence of a winning strategy for a reduction game (I.4.2.9.). Chapter I is devoted to the general results used in the sequel and to the explicit formulation of the main result I.4.2.9.. If r = adapted order, it is enough to obtain adapted order $\leq r-1$ and to continue so. Chapters II-V deal with the various types of vector fields which fall into the preceding ones after a finite number of steps of the process. In II, the more stable case is treated. The control of this case is made in a very similar way to the control of surface singularities by means of the Newton polygon (see |10|). III and IV deal with the reduction game beginning at the so-called "type one" and the most complicated computations are contained there. The remaining cases are considered in V. In general, the classification of the various types is made by considering the transversality positions between the exceptional divisor and some ideals associated to the vector field which play a role similar to the role of the strict tangent space in the case of surface singularities (see III.1).

I wish to express my gratitude to Professor J.M. Aroca for introducing me to the subject, for his constant aid and for guiding me during the realization of this work. To Professor J. Giraud for his interest and for his technical aid, essential for the precise formulation of many parts of these notes. To Professor A. Campillo for his important suggestions. Special thanks to Miss A. Artero for her hard typing task. (*)

(*) This work has been partially supported by the CAICYT.

-CONTENTS-

RESOLUTION STATEMENTS FOR A VECTOR FIELD

0. INTRODUCTION

In this chapter we shall develop the preliminary concepts and results that we shall need in the formulation of the main statements about resolution games to which proof this work is devoted. We shall also formulate more general statements about resolution of vector fields that we shall not prove.

1. ADAPTED VECTOR FIELDS

(1.1) General hypothesis and notations

(1.1.1) In the sequel, X will denote a regular variety, i.e. a regular integral separated scheme of finite type over an algebraically closed field k. Let us denote by n the dimension of X.

(1.1.2) Remark. Along this chapter we shall not make any assumption on the dimension n nor on the characteristic of k. But in the following chapters we shall assume that n = 3 and that the characteristic of k is zero. We would like to remark that although the assumption on the dimension is essential for the good work of the techniques used, it is not so for the assumption on the characteristic of k. Actually most of the results remain true for positive characteristic and we shall always indicate where the hypothesis on zero characteristic is used.

(1.1.3) We shall denote by Ω_X, resp. Ξ_X, the cotangent sheaf, resp. the tangent sheaf, of X relatively to k. Because of our assumptions on X, both Ω_X and Ξ_X are locally free $\mathcal{O}_X$-modules of rank n.

(1.1.4) We shall say that a closed subscheme E of X is a "normal crossings divisor" iff for each closed point P of X there is a regular system of parameters $x_1,\dots,x_n$ (r.s. of p. for short) of the local ring $\mathcal{O}_{X,P}$ of P in X such that for some s, $0\leq s\leq n$, the ideal of E at P is generated by $x_1.x_2\dots x_s$.

(1.1.5) Let E be a normal crossings divisor and let $\mathcal{J}_E$ be the sheaf of ideals of E. We shall denote by $\Xi_X[E]$ the (unique) subsheaf of Ξ_X such that for every point P of X one has that

(1.1.5.1) $$\Xi_{X,P}[E] = \{D \in \Xi_{X,P};\ D(\mathcal{J}_{E,P}) \subset \mathcal{J}_{E,P}\}$$

(recall that $\Xi_{X,P}$ is the module of k-derivations of the local ring $\mathcal{O}_{X,P}$).

(1.1.6) Let P be a closed point of X, let E be a normal crossings divisor and let $x=(x_1,\dots,x_n)$ be a regular system of parameters of $\mathcal{O}_{X,P}$ such that E is given at P by $x_1\cdot\dots\cdot x_s$. Let $\partial/\partial x_i$, i=1,...,n, be the k-derivations of $\mathcal{O}_{X,P}$ given by

(1.1.6.1) $$\partial/\partial x_i(x_j) = \delta_{ij} \qquad \text{(Kronecker symbol).}$$

Then standard computations show that $\partial/\partial x_i$, i=1,...,n, is a free basis of $\Xi_{X,P}$ and that

(1.1.6.2) $$x_1\partial/\partial x_1,\dots,x_s\partial/\partial x_s,\ \partial/\partial x_{s+1},\dots,\partial/\partial x_n$$

is a free basis of $\Xi_{X,P}[E]$.

(1.2) Vector fields and distributions

(1.2.1) Definition. Let P be a closed point of X. We shall call "vector field at P" to any element D of $\Xi_{X,P}$. Any inversible $\mathcal{O}_X$-submodule of Ξ_X will be called "an unidimensional distribution over X".

(1.2.2) For any unidimensional distribution $\mathcal{D}$ and any closed point P, the stalk $\mathcal{D}_P$ is generated by a vector field D. If D' generates $\mathcal{D}_P$ there is a unit u in $\mathcal{O}_{X,P}$ such that D' = u.D.

(1.2.3) Proposition. Let us consider the natural pairing

(1.2.3.1) $$<,>: \Omega_X \times \Xi_X \longrightarrow \mathcal{O}_X$$

and for a given unidimensional distribution $\mathcal{D}$ let us denote by $\alpha(\mathcal{D})$ the double orthogonal of $\mathcal{D}$ relatively to <,>. Then

a) $\alpha(\mathcal{D})$ is an unidimensional distribution over X.

b) Let P be any closed point of X and let D be a generator of $\mathcal{D}_P$ such that

(1.2.3.2) $$D = \sum_{i=1,\dots,n} a_i \partial/\partial x_i, \quad a_i \in \mathcal{O}_{X,P};$$

where $x = (x_1,\dots,x_n)$ is a r.s. of p. of $\mathcal{O}_{X,P}$. Let b be the g.c.d. of a_i, i=1,..,n. Then $\alpha(\mathcal{D})_P$ is generated by D/b.

Proof It is enough to remark that for a closed point P

(1.2.3.3) $$\alpha(\mathcal{D})_P = \{D' \in \Xi_{X,P};\ \omega \in \Omega_{X,P},\ <\omega, D''>=0\ \forall D'' \in \mathcal{D}_P \Rightarrow <\omega,D'>=0\}$$

that both $\Omega_{X,P}$ and $\Xi_{X,P}$ are free $\mathcal{O}_{X,P}$-modules, that

(1.2.3.4) $$\Xi_{X,P} = \mathrm{Hom}_{\mathcal{O}_{X,P}}(\Omega_{X,P},\mathcal{O}_{X,P})$$

and that $\mathcal{O}_{X,P}$ is a U.F.D. because it is regular.

(1.2.4) Definition. We shall say that an unidimensional distribution $\mathcal{D}$ is "multiplicatively irreducible" iff $\mathcal{D} = \alpha(\mathcal{D})$ and we shall call $\alpha(\mathcal{D})$ the "multiplicative reduction of $\mathcal{D}$". (For short, m.i.u.d.= multiplicatively irreducible unidimensional distribution).

(1.3) The adapted case

(1.3.1) Let E be a normal crossings divisor on X.

(1.3.2) Definition. Let P be a closed point of X. We shall say that a vector field D at P is "adapted to E" iff $D \in \Xi_{X,P}[E]$. A unidimensional distribution $\mathcal{D}$ over X will be called "adapted to E" iff $\mathcal{D} \subset \Xi_{X,P}[E]$.

(1.3.3.) Proposition. Let $\mathcal{D}$ be an unidimensional distribution over X, then

a) The $\mathcal{O}_X$-submodule $(\mathcal{D},E)$ given by

(1.3.3.1)
$$(\mathcal{D},E) = \mathcal{D} \cap \Xi_X[E]$$

is an adapted to E unidimensional distribution over X.

b) Let P be a closed point of X, let $x = (x_1,\ldots,x_n)$ be a r.s. of p. of $\mathcal{O}_{X,P}$ such that E is given at P by $x_1 \ldots x_s$ and let

(1.3.3.2)
$$D = \sum_{i=1,\ldots,n} a_i \partial/\partial x_i$$

be a generator of $\mathcal{D}_P$. Then $(\mathcal{D},E)_P$ is generated by

(1.3.3.3)
$$x_1^{\varepsilon_1} \ldots x_s^{\varepsilon_s} . D$$

where $\varepsilon_j = 1$ if $a_j \notin x_j.\mathcal{O}_{X,P}$, $\varepsilon_j = 0$ otherwise, $j=1,\ldots,s$.

Proof It is enough to remark that for a closed point P of X one has that

(1.3.3.4)
$$(\mathcal{D},E)_P = \{\lambda.D;\ \lambda \in I\}$$

where I is the principal ideal of $\mathcal{O}_{X,P}$ given by

(1.3.3.5) $$I = \{ \lambda \in \mathcal{O}_{X,P};\ \lambda . D(x_1,\dots,x_s) \in (x_1 \dots x_s)\mathcal{O}_{X,P} \}.$$

(1.3.4) Definition. We shall call the unidimensional distribution $(\mathcal{D},E)$ of (1.3.3.1) to be the "adaptation of $\mathcal{D}$ to E".

(1.3.5) Let $\mathcal{D}$ be an adapted to E unidimensional distribution and let $\Xi_X[E]^\wedge$ denote the dual sheaf of $\Xi_X[E]$. Let us denote by $\alpha'(\mathcal{D},E)$ the double orthogonal of $\mathcal{D}$ with respect to the natural pairing

(1.3.5.1) $$\langle , \rangle_E : \Xi_X[E] \times \Xi_X[E]^\wedge \longrightarrow \mathcal{O}_X.$$

Then computations like (1.2.3) show that $\alpha'(\mathcal{D},E)$ is an adapted to E unidimensional distribution. Moreover, if P is a closed point of X, a generator of $\alpha'(\mathcal{D},E)_P$ can be obtained from a generator of $\mathcal{D}_P$ by dividing it by the g.c.d. of its coefficients in any basis of $\Xi_{X,P}[E]$.

Now, combining (1.2.3) b) and (1.3.3) b) one can deduce that

(1.3.5.2) $$\alpha'(\mathcal{D},E) = (\alpha(\mathcal{D}),E).$$

(1.3.6) Definition. Let $\mathcal{D}$ be an adapted to E unidimensional distribution. We shall say that $\mathcal{D}$ is "multiplicatively irreducible relative to E" iff

(1.3.6.1) $$\mathcal{D} = (\alpha(\mathcal{D}),E)$$

And we shall call $(\alpha(\mathcal{D}),E)$ the "multiplicative reduction relative to E" of $\mathcal{D}$.

(1.3.7) Remark. The results of this section are true for the case "E is a closed subscheme of X of pure codimension one".

(1.4) The formal case

(1.4.1) Let (R,M,k) be a complete local regular ring of dimension n containing k (actually R is a ring of formal power series over k). Let $\hat{X}$ be the scheme Spec (R). Let $\hat{E}$ be a normal crossings divisor of $\hat{X}$ as in (1.1.4) given at the closed point of $\hat{X}$ by the ideal $I \subset R$. Then one has that both

(1.4.1.1) $$\mathrm{Der}_k(R)$$

and

(1.4.1.2) $$\mathrm{Der}_k(R)[I] = \{ D \in \mathrm{Der}_k(R);\ D(I) \subset I\}$$

are free R-modules of rank n.

(1.4.2) Definition. With notations as above, we shall call "formal vector fields over the closed point of $\hat{X}$" (resp. "and adapted to $\hat{E}$") the elements of $\mathrm{Der}_k(R)$ (resp. of $\mathrm{Der}_k(R)[I]$). We shall call "formal unidimensional distribution over the closed point of $\hat{X}$" (resp. "and adapted to $\hat{E}$") the rank-1 free R-submodules of $\mathrm{Der}_k(R)$ (resp. of $\mathrm{Der}_k(R)[I]$).

(1.4.3) Let $\mathcal{D}$ be a formal unidimensional distribution over the closed point of $\hat{X}$ We shall denote the double orthogonal of $\mathcal{D}$ in $\mathrm{Der}_k(R)$ by $\alpha(\mathcal{D})$ and we shall denote $(\mathcal{D},\hat{E}) = \mathcal{D} \cap \mathrm{Der}_k(R)[I]$ Like in (1.3.6), we shall say that $\mathcal{D}$ is "multiplicatively irreducible relatively to $\hat{E}$" iff $\mathcal{D} = (\alpha(\mathcal{D}),\hat{E})$.

(1.4.4) Let P be a closed point of X and $R = \hat{\mathcal{O}}_{X,P}$ (the completion of the local ring $\mathcal{O}_{X,P}$). Then any k-derivation $D \in \Xi_{X,P}$ may be uniquely extended to a k-derivation $\hat{D} \in \mathrm{Der}_k(R)$. If E is a normal crossings divisor of X given at P by $I \subset \mathcal{O}_{X,P}$, then any vector field $D \in \Xi_{X,P}[E]$ be extended to $\hat{D} \in \mathrm{Der}_k(R)[\hat{I}]$, where $\hat{I} = I.R$. We shall call $\hat{D}$ "the associated to D formal vector field". The same procedure can be used for an unidimensional distribution in order to obtain the "associated to $\mathcal{D}$ formal distribution $\hat{\mathcal{D}}_P$ at P".

2. BLOWING-UPS OF VECTOR FIELDS

(2.1) Inverse images by a morphism

(2.1.1) Let $\pi : X' \longrightarrow X$ be a morphism between n-dimensional regular varieties over k, and let $\mathcal{D}$ be an $\mathcal{O}_X$-submodule of Ξ_X. Let $\mathcal{D}'$ be the image of $\mathcal{D}$ by the natural morphism of $\mathcal{O}_X$-modules

(2.1.1.1) $$\phi : \Xi_X \longrightarrow \mathrm{Hom}_{\mathcal{O}_X}(\Omega_X, \pi_* \mathcal{O}_{X'})$$

induced by the structural morphism $\pi^{\#} : \mathcal{O}_X \longrightarrow \pi_* \mathcal{O}_{X'}$. Because of the finiteness of Ω_X over $\mathcal{O}_X$ and the adjoint property of the inverse-direct image, one has a natural isomorphism of $\mathcal{O}_X$-modules

(2.1.1.2) $$\eta : \mathrm{Hom}_{\mathcal{O}_X}(\Omega_X, \pi_* \mathcal{O}_{X'}) \longrightarrow \longrightarrow \pi_* \mathrm{Hom}_{\mathcal{O}_{X'}}(\pi^* \Omega_X, \mathcal{O}_{X'}).$$

Let $\mathcal{D}''$ be the image of $\mathcal{D}'$ by η. Now, let $\mathcal{D}'''$ be the image of the natural morphisms

(2.1.1.3) $$\pi^* \mathcal{D}'' \longrightarrow \pi^* \pi_* \mathrm{Hom}_{\mathcal{O}_{X'}}(\pi^* \Omega_X, \mathcal{O}_{X'}) \longrightarrow \longrightarrow \mathrm{Hom}_{\mathcal{O}_{X'}}(\pi^* \Omega_X, \mathcal{O}_{X'}).$$

Finally, let us denote by $\mathcal{D}^{\pi}$ the inverse image of $\mathcal{D}'''$ by the natural morphism of $\mathcal{O}_{X'}$-modules

(2.1.1.4) $$\Xi_{X'} \longrightarrow \mathrm{Hom}_{\mathcal{O}_{X'}}(\pi^* \Omega_X, \mathcal{O}_{X'})$$

obtained by applying the $\mathrm{Hom}_{\mathcal{O}_{X'}}(- , \mathcal{O}_{X'})$ functor to the natural exact sequence

(2.1.1.5) $$\pi^* \Omega_{X'} \longrightarrow \Omega_X \longrightarrow \Omega_{X'/X} \longrightarrow 0.$$

(2.1.2) <u>Definition</u>. With notations as above, $\mathcal{D}^{\pi}$ will be called the inverse image of $\mathcal{D}$ by π.

(2.1.3) Let us suppose that X' = Spec (A'), that X = Spec (A) and that $\mathcal{D}$ is a coherent $\mathcal{O}_X$-submodule of Ξ_X. Because of Ω_X and $\Omega_{X'}$ are of finite type, Ξ_X and $\Xi_{X'}$ are coherent modules given respectively by

(2.1.3.1) $$\mathrm{Der}_k(A)^\sim \text{ and } \mathrm{Der}_k(A')^\sim.$$

("~" means "associated module over the scheme"). Let us suppose that $\mathcal{D}$ is given by $M^\sim$ where M is an A-submodule of $\mathrm{Der}_k(A)$. Let M' be the image of the A-morphism

(2.1.3.2) $$\mathrm{Der}_k(A) \longrightarrow \mathrm{Hom}_A(\Omega_A, A').$$

Now, since Ω_A is locally free one has that

(2.1.3.2) $$\mathrm{Hom}_A(\Omega_A, A') = \mathrm{Hom}_{A'}(\Omega_A \otimes_A A', A') \cong$$
$$\cong \mathrm{Hom}_{A'}(\Omega_A \otimes_A A', A') \otimes_A A'.$$

Let M''' be the image of $M' \otimes_A A'$ by

(2.1.3.3) $$M' \otimes_A A' \longrightarrow \mathrm{Hom}_{A'}(\Omega_A \otimes_A A', A').$$

And finally, if M^π is the inverse image of M''' by the natural morphism of A'-modules

(2.1.3.4) $$\mathrm{Der}_k(A') \longrightarrow \mathrm{Hom}_{A'}(\Omega_A \otimes_A A', A').$$

Then one has that $\mathcal{D}^\pi$ is a coherent $\mathcal{O}_{X'}$-submodule of $\Xi_{X'}$ given by $(M^\pi)^\sim$.

(2.1.4) Proposition. Let $\pi : X' \longrightarrow X$ be a birrational morphism and let $\mathcal{D}$ be an unidimensional distribution over X. Then $\mathcal{D}^\pi$ is also an unidimensional distribution.

Proof Let P' be a closed point of X' and P = π(P'). Then, with notation like in (2.1.1) one has that

(2.1.4.1) $$(\pi^*\mathcal{D}'')_{P'} = \mathcal{D}''_P \otimes_{\mathcal{O}_{X,P}} \mathcal{O}_{X',P'}.$$

It follows that $\mathcal{D}''_{P'}$ is generated by a single element. Since π is birrational, $\Omega_{X'/X,P'}$ is a torsion $\mathcal{O}_{X',P'}$-module and

(2.1.4.2) $$0 \longrightarrow \Xi_{X',P'} \longrightarrow [\mathrm{Hom}_{\mathcal{O}_{X'}}(\pi^*\Omega_X, \mathcal{O}_{X'})]_{P'}$$

is exact. Now, both $\Xi_{X',P'}$ and the right hand side in (2.1.4.2) are free $\mathcal{O}_{X',P'}$ modules of rank n. Since $\mathcal{O}_{X',P'}$ is a U.F.D. one deduces that $\mathcal{D}^{\pi}_{P'}$ is generated by a single element. Now the result follows from the fact that $\mathcal{D}^{\pi}$ is coherent in view of (2.1.3).

(2.2) Adapted blowing-ups

(2.2.1) Let E be a normal crossings divisor of X and let Y be a closed subscheme of X. Y has "normal crossings" with E iff the following condition is satisfied "for each closed point P of X there exists a regular system of parameters $x = (x_1,\ldots,x_n)$ of $\mathcal{O}_{X,P}$ and two sets A,B contained in $\{1,\ldots,n\}$ such that

(2.2.1.1)
$$I_{E,P} = \left(\prod_{i \in A} x_i\right).\mathcal{O}_{X,P}$$
$$I_{Y,P} = \sum_{i \in B} x_i.\mathcal{O}_{X,P}$$

where $I_{E,P}$ and $I_{Y,P}$ denote respectively the ideals of E and Y in $\mathcal{O}_{X,P}$. We shall say that a r.s. of p. satisfying the conditions (2.2.1.1) is "suited for the pair (E,Y)". If $P \notin E$ we assume $B = \emptyset$.

(2.2.2) Let E,Y be as above, let us suppose that Y has normal crossings with E, let us denote by

(2.2.2.1) $$\pi : X' \longrightarrow X$$

the blowing-up with center Y and let us denote by E' the closed subscheme (with reduced structure) given by $\pi^{-1}(Y \cup E)$. Then E' is a normal crossings divisor of X'.

(2.2.3) Definition. Let $\mathcal{D}$ be a multiplicatively irreducible unidimensional distribution over X (resp. and adapted to E). We shall say that $\alpha(\mathcal{D}^{\pi})$ is the "strict transform of $\mathcal{D}$ by π", resp. that $(\alpha(\mathcal{D}^{\pi}),E')$ is the "strict transform of $\mathcal{D}$ by π adapted to E".

(2.2.4) Remark. Let $\mathcal{U} = X-Y$, $\mathcal{U}' = \pi^{-1}(\mathcal{U})$. Since $\pi|_{\mathcal{U}'}: \mathcal{U}' \longrightarrow \mathcal{U}$ is an isomorphism and $E' \cap \mathcal{U}'$ and $E \cap \mathcal{U}$ correspond one to another by $\pi|_{\mathcal{U}'}$, one has that $\mathcal{D}^{\pi}|_{\mathcal{U}'} \cong \mathcal{D}|_{\mathcal{U}}$ by the induced isomorphism between the tangent sheaves of $\mathcal{U}'$ and $\mathcal{U}$. Then one has that

$$(2.2.4.1) \qquad (\alpha(\mathcal{D}^{\pi}),E')|_{\mathcal{U}'} = \mathcal{D}^{\pi}|_{\mathcal{U}'}.$$

So, the difference between $(\mathcal{D}^{\pi},E')$ and $(\alpha(\mathcal{D}^{\pi}),E')$ is concentrated in $\pi^{-1}(Y)$.

Let P' be a closed point of $\pi^{-1}(Y)$ and let f be a local equation of $\pi^{-1}(Y)$ at P'. Let D' be a generator of $(\alpha(\mathcal{D}^{\pi}),E')_{P'}$ and let D" be a generator of $(\mathcal{D}^{\pi},E')_{P'}$. In view of (1.3.3.3), (1.3.5) and (2.2.4.1) one has that D' and D" may be chosen in such a way that there exists an integer μ such that

$$(2.2.4.2) \qquad f^{\mu}.D' = D''.$$

This number μ does not depend on the closed point P'. We shall call it the "blowing-up order of π on Y adapted to E" and we shall denote it by

$$(2.2.4.3) \qquad \mu(\mathcal{D},E;Y).$$

(2.2.5) Equations. Let us consider the situation of (2.2.1) and (2.2.2) and let P' be a closed point of X' such that $\pi(P') = P$. Let us denote $S=\mathcal{O}_{X,P}$ and $S'=\mathcal{O}_{X',P'}$. Then $S \subset S'$, k is a coefficient field for S' and there exist a r.s. of p. $x'_1,\ldots,x'_n$ of S', $i_o \in B$ and scalars $\zeta_i \in k$, $i \in B - \{i_o\}$, such that

$$(2.2.5.1) \qquad \begin{aligned} x_{i_o} &= x'_{i_o} \\ x_i &= (x'_i + \zeta_i)x'_{i_o} \qquad i \in B - \{i_o\} \\ x_i &= x'_i \end{aligned}$$

Let us consider the natural mapping

$$(2.2.5.2) \qquad \phi: \Omega_S \otimes_S S' \longrightarrow \Omega_{S'}.$$

The left hand side is a rank-n free S'-module generated by $dx_i \otimes 1$, $i=1,\ldots,n$ and one has that

$$(2.2.5.3)\qquad \phi(dx_i \otimes 1) = dx_i = \begin{cases} dx'_{i_o} & \text{if } i = i_o \\ (x'_i+\zeta_i)dx'_{i_o} + x'_{i_o}dx_i & \text{if } i \in B - \{i_o\} \\ dx'_i & \text{if } i \notin B \end{cases}$$

Thus, if we denote by ∂_i, $i=1,\dots,n$ the dual basis of $(dx_i \otimes 1)$, then the induced morphism

$$(2.2.5.4)\qquad \psi : Der(S') \longrightarrow Hom_{S'}(\Omega_S \otimes_S S', S')$$

is given by

$$(2.2.5.5)\qquad \psi(\partial/\partial x'_i) = \begin{cases} \partial_{i_o} + \sum_{j \in B-\{i_o\}} (x'_j+\zeta_j)\partial_j & i = i_o \\ x'_{i_o} \cdot \partial_i & i \in B-\{i_o\} \\ \partial_i & i \notin B \end{cases}$$

Now, let us consider the set A' given by

$$(2.2.5.6)\qquad A' = \{i_o\} \cup \{i \in A \cap B;\ \zeta_i = 0\} \cup (A-B).$$

Then one has that E' is given at P' by

$$(2.2.5.7)\qquad \prod_{j \in A'} x'_j = 0$$

Remark that A', and so E', do not depend on the choice of i_o made in (2.2.5.1).

Now, let us suppose that $\mathcal{D}_P$ is generated by

$$(2.2.5.8)\qquad D = \sum_{i \in A} a_i x_i \partial/\partial x_i + \sum_{i \notin A} a_i \partial/\partial x_i .$$

Then, the corresponding module in $Hom_{S'}(\Omega_S \otimes S', S')$ is generated by

$$(2.2.5.9)\qquad D'' = \sum_{i \in A} a_i x_i \partial_i + \sum_{i \notin A} a_i \partial_i .$$

Now, in view of (2.2.5.5) one can obtain a generator of $(\mathcal{D}^{\pi},D')_{P'}$. We shall distinguish two cases: $i_o \in A$, $i_o \notin A$. If $i_o \in A$, then $\mathcal{D}^{\pi}_{P'}$ is generated by

(2.2.5.10)
$$D' = [x'_{i_o}]^{\varepsilon}[a_{i_o} x'_{i_o} \partial/\partial x'_{i_o} + \sum_{i \in A' \cap B-\{i_o\}} (a_i - a_{i_o}) x'_i \partial/\partial x'_i + \sum_{i \in (A-A') \cap B-\{i_o\}} (a_i - a_{i_o})(x'_i + \zeta_i)\partial/\partial x'_i + \sum_{i \in B-A} (a_i/x'_{i_o} - (x'_i+\zeta_i)a_{i_o})\partial/\partial x'_i + \sum_{i \in A'-B} a_i x'_i \partial/\partial x'_i + \sum_{i \notin A' \cup B} a_i \partial/\partial x'_i].$$

where $\varepsilon = 1$ iff there is a_i, $i \in B-A$, such that $a_i \notin x'_{i_o}.S'$ and $\varepsilon = 0$ otherwise. In this case D' is adapted to E' and it is a generator of $(\mathcal{D}^{\pi},E')_{P'}$. Let us suppose that $i_o \notin A$, then $(\mathcal{D}^{\pi},E')_{P'}$ is generated by

(2.2.5.11)
$$D' = [x'_{i_o}]^{\varepsilon}[(a_{i_o}/x'_{i_o}).x'_{i_o}.\partial/\partial x'_{i_o} + \sum_{i \in A' \cap B-\{i_o\}} (a_i - a_{i_o}/x'_{i_o}) x'_i . \partial/\partial x'_i + \sum_{i \in (A-A') \cap B} (a_i - a_{i_o}/x'_{i_o})(x'_i+\zeta_i)\partial/\partial x'_i + \sum_{i \in B-A-\{i_o\}} (a_i/x'_{i_o} - (x'_i+\zeta_i)\, a_{i_o}/x'_{i_o})\partial/\partial x'_i + \sum_{i \in A'-B} a_i x'_i \partial/\partial x'_i + \sum_{i \notin A' \cup B} a_i \partial/\partial x'_i].$$

where $\varepsilon = 1$ iff there is a_i, $i \in B-A$ (note $i_o \in B-A$) such that $a_i \notin x'_{i_o}.S'$ and $\varepsilon = 0$ otherwise. Remark that in this case one can have that D'/x'_{i_o} is a generator of $\mathcal{D}^{\pi}_{P'}$, if, for instance, one has that $A = \emptyset$, $B = \{1,\dots,n\}$ and the only $a_i \notin x'_{i_o} S'$ is a_{i_o}.

In both cases, and in view of (2.2.4), the strict transform $(\alpha(\mathcal{D}^{\pi}),E')_{P'}$ is generated by

(2.2.5.12)
$$[x'_{i_o}]^{-\mu}.D'$$

where $\mu = \mu(\mathcal{D},E;Y)$.

(2.3) Formal blowing-ups

(2.3.1) Let us adopt the notation of (1.4.1). For a regular closed subscheme $Y^\wedge$ of $X^\wedge$ one can establish the concept of "normal crossings with $E^\wedge$" and of "r.s. of p. suited for the pair $(E^\wedge,Y^\wedge)$" in the same way as in (2.2.1). Now, let $\mathcal{D}^\wedge$ be a formal multiplicatively irreducible relatively to $E^\wedge$ unidimensional distribution over the closed point P of $X^\wedge$.

(2.3.2) Let π: $X' \longrightarrow X^\wedge$ be the blowing-up of $X^\wedge$ with center $Y^\wedge$ and let P' be a closed point of $\pi^{-1}(Y^\wedge)$ such that $\pi(P') = P$. Let $R' = \mathcal{O}_{X',P'}{}^\wedge$. We shall call the morphism

(2.3.2.1) $$\pi^\wedge\colon \operatorname{Spec}(R') \longrightarrow X^\wedge$$

"directional blowing-up of $X^\wedge$ with center $Y^\wedge$ in the direction of P'".

(2.3.3) The morphism (2.3.2.1) corresponds to a ring morphism $R \to R'$ between complete regular local rings having k as a coefficient field. One has that the R'-module $\operatorname{Der}_k(R,R')$ is free of rank n and it is isomorphic to $\operatorname{Der}_k(R) \otimes_R R'$. Let $\mathcal{D}^{\wedge\prime}$ be the image of the morphism

(2.3.3.1) $$\mathcal{D}^\wedge \otimes_R R' \longrightarrow \operatorname{Der}_k(R,R')$$

and let $\mathcal{D}^\wedge$ be the inverse image of $\mathcal{D}^{\wedge\prime}$ by the injection (given by restriction)

(2.3.3.2) $$\operatorname{Der}_k(R') \longrightarrow \operatorname{Der}_k(R,R')$$

Then $\mathcal{D}^\wedge$ is a formal unidimensional distribution over the closed point of $X^{\wedge\prime} = \operatorname{Spec}(R')$. It will be called "inverse image of $\mathcal{D}^\wedge$ by $\pi^\wedge$".

(2.3.4) Definition. With notations as above, we shall call $\alpha(\mathcal{D}^{\pi\wedge})$ the "strict transform of $\mathcal{D}^\wedge$ by $\pi^\wedge$". If $E^{\wedge\prime} = \pi^{\wedge -1}(E^\wedge \cup Y^\wedge)$, we shall call $(\alpha(\mathcal{D}^{\pi\wedge}),E^{\wedge\prime})$ the "adapted to $E^\wedge$ strict transform of $\mathcal{D}^\wedge$ by $\pi^\wedge$".

(2.3.5) Equations. Let $x_1,\ldots,x_n$ be a r.s. of p. of R suited for the pair $(E^\wedge,Y^\wedge)$. Then there is a r.s. of p. of R' $x'_1,\ldots,x'_n$ such that (2.2.5.1) holds. Let us denote by ∂_i, i=1,...,n, the R'-basis of $\mathrm{Der}_k(R,R')$ given by $\partial_i = \partial/\partial x_i \otimes 1$. Then (2.3.3.3) has just the expression of (2.2.5.5) and the equations for $(\mathcal{D}^\wedge,E^{\wedge\prime})$ and $(\alpha(\mathcal{D}^{\pi\wedge}),E^{\wedge\prime})$ can be deduced exactly in the same way as in (2.2.5).

(2.3.6) Proposition. Let $X,\mathcal{D},E,Y,P$ and $\pi: X' \longrightarrow X$ be as in (2.2). Let $\mathcal{D}^\wedge = \mathcal{D}_P^\wedge$ be the associated to $\mathcal{D}$ formal distribution at P. Let $X^\wedge$ be Spec $(\mathcal{O}^\wedge_{X,P})$ and $E^\wedge,Y^\wedge$ the corresponding subschemes. Let $\pi^*: X^* \longrightarrow X^\wedge$ be the blowing-up with center $Y^\wedge$. Let P' be a closed point of X such that $\pi(P') = P$ and let P* be the closed point of X^* associated to P' by the universal property of the blowing-up. Let $X^{\wedge\prime}=\mathrm{Spec}(\mathcal{O}^\wedge_{X^*,P^*}) = \mathrm{Spec}(\mathcal{O}^\wedge_{X',P'})$ and let $\pi^\wedge: X^{\wedge\prime}\ \ X^\wedge$ be the corresponding morphism. Then one has that

$$\mathcal{D}^{\pi\wedge} = (\mathcal{D}^\pi)^\wedge_{P'}$$

(2.3.6.1)
$$(\mathcal{D}^{\pi\wedge},E^{\wedge\prime}) = (\mathcal{D}^\pi,E')^\wedge_{P'}$$

$$(\alpha(\mathcal{D}^{\pi\wedge}),E^{\wedge\prime}) = (\alpha(\mathcal{D}^\pi),E')^\wedge_{P'}$$

Proof. It follows from (2.2.5) and (2.3.5).

3. SINGULAR LOCUS AND BLOWING-UP

(3.1) Adapted order of a vector field

(3.1.1) Definition. Let $\mathcal{D}$ be an unidimensional distribution over X which is adapted to a n.c. divisor E. Let Q be a (not necessarily closed) point of X. The "adapted order of $\mathcal{D}$ at Q" will be the maximum of the integers m such that

(3.1.1.1)
$$\mathcal{D}_Q \subset \eta^m \cdot \Xi_{X,Q}[E]$$

where η is the maximal ideal of $\mathcal{O}_{X,Q}$. It will be denoted by $\nu(\mathcal{D},E,Q)$. If

$Z \subset X$ is a closed subscheme and Q its generic point, we shall denote $\nu(\mathcal{D},E,Z) = \nu(\mathcal{D},E,Q)$. (For short: n.c.= normal crossings).

(3.1.2) In the formal case, one proceeds in a similar way. Let $X^\wedge$ be the scheme Spec (R) where R is a complete local regular ring having k as a coefficient field, let $E^\wedge$ be a n.c. divisor on $X^\wedge$ and let $\mathcal{D}^\wedge$ be a formal unidimensional distribution over the closed point of $X^\wedge$ which is adapted to $E^\wedge$. Let Q be a point of $X^\wedge$ and η the corresponding ideal of R. Then $\nu(\mathcal{D}^\wedge,E^\wedge,Q)$ will be the maximum integer m such that

(3.1.2.1) $$\mathcal{D}^\wedge \subset \eta^m.\ \mathrm{Der}_k\ (R)[I]$$

where I = ideal of E.

(3.1.3). With notations as above, let P be a closed point of X such that $P \in \overline{\{Q\}}$ and let $x_1,\ldots,x_n$ be a r.s. of p. of $\mathcal{O}_{X,P}$ suited for $(E,\{P\})$ such that E is given by $\prod_{i \in A} x_i$. Let us suppose that $\mathcal{D}_P$ is generated by

(3.1.3.1) $$D = \sum_{i \in A} a_i x_i \partial/\partial x_i + \sum_{i \notin A} a_i \partial/\partial x_i .$$

Let η be the ideal of Q in $\mathcal{O}_{X,P}$. Then one has that

(3.1.3.2) $$\nu(\mathcal{D},E,Q) = \min\ (\nu_\eta(a_i);\ i=1,\ldots,n)$$

where $\nu_\eta(a_i)$ is the η-adic order of a_i. Moreover, if $Q^\wedge,\mathcal{D}^\wedge,\ldots$ are the correspon-ding formal objects, one has that

(3.1.3.3) $$\nu(\mathcal{D}^\wedge_P,E^\wedge,Q^\wedge) = \nu(\mathcal{D},E,Q).$$

Finally, one deduces easily an expression like (3.1.3.2) for the formal case.

(3.1.4) <u>Theorem</u>. Let P be a closed point of X and let $\pi: X' \longrightarrow X$ be the blowing-up with center P. Then, for any adapted to E multiplicatively irreducible unidimensio-nal distribution $\mathcal{D}$ over X and for any closed point P' of X' such that $\pi(P') = P$ one has that

(3.1.4.1) $$\nu(\mathcal{D},E,P) \geq \nu((\alpha(\mathcal{D}^{\pi}),E'),E',P').$$

Proof. Let $x_1,\dots,x_n$ be a r.s. of p. of $\mathcal{O}_{X,P}$ suited for (E,P) and let us suppose that $\mathcal{D}_P$ is generatd by

(3.1.4.2) $$D = \sum_{i \in A} a_i x_i \partial/\partial x_i + \sum_{i \notin A} a_i \partial/\partial x_i .$$

Let us denote $r = \nu(\mathcal{D},E,P)$, $r' = \nu((\alpha(\mathcal{D}^{\pi}),E'),E',P')$. Considering the equations (2.2.5.10), (2.2.5.11) and (2.2.5.12), there are two posibilites: $\varepsilon = 0$ or $\varepsilon = 1$. If $\varepsilon = 1$ then $r = r' = 0$. If $\varepsilon = 0$, there are two posibilities for $\mu = \mu(\mathcal{D},E,P)$: $\mu = r$ or $\mu = r-1$. If $\mu = r$ then there exist $i \in A$ such that $\nu(a_i) = r$, where $\nu(a_i)$ is the order of a_i with respect to the η_P-adic filtration of $\mathcal{O}_{X,P}$. Then the result follows from the equations. If $\mu = r-1$, then there exist $i \notin A$ such that $\nu(a_i) = r$ and the result follows in an analog way.

(3.1.5) Corollary. Let X^ be as in 2.3 and let π^: X^' ⟶ X^ be the directional blowing-up of X^ with center in the closed point P of X^ and in the direction of P'. Let $\mathcal{D}$^ be a formal m. i. u. d. defined over the closed point of X^. Then one has that

(3.1.5.1) $$\nu(\hat{\mathcal{D}},\hat{E},P) \geq \nu((\alpha(\mathcal{D}^{\hat{\pi}}),\hat{E}'),\hat{E}',P').$$

Proof. The same one as in (3.1.4).

(3.1.6) Remark. Stability results as (3.1.4) and (3.1.5) do not work for the non adapted case. For instance, if $\mathcal{D}_P$ is generated by

(3.1.6.1) $$D = y\partial/\partial x + x^3\partial/\partial y$$

and we made $x = x'$, $y = x'y'$, the strict tansform is generated by

(3.1.6.2) $$D' = y'x'\partial/\partial x' + (x'^2 - y'^2)\partial/\partial y'$$

and the (non adapted) order has been increased by a unit.

(3.1.7) Definition. Let $\mathcal{D}$ be a m. i. u. d. defined over

X, and adapted to E. "The singular locus of $\mathcal{D}$ relatively to E", denoted by Sing $(\mathcal{D},E)$, will be the set of points Q of X such that $\nu(\mathcal{D},E,Q) \geq 1$. For any $r \geq 0$, we shall denote

(3.1.7.1) $$\mathrm{Sing}^r(\mathcal{D},E) = \{ Q \in X;\ \nu(\mathcal{D},E,Q) \geq r \}.$$

(3.1.8) The adapted order is semicontinuous. So Sing $(\mathcal{D},E)$ and $\mathrm{Sing}^r(\mathcal{D},E)$ are closed in X. Moreover, the fact that $\mathcal{D}$ is multiplicatively irreducible is equivalent to say that Sing $(\mathcal{D},E)$ has no components of codimension one. Finally, because of the noetherian properties of X, there is a maximum r such that $\mathrm{Sing}^r(\mathcal{D},E) \neq \emptyset$, we shall denote by Sam $(\mathcal{D},E)$ this $\mathrm{Sing}^r(\mathcal{D},E)$.

(3.1.9) The concepts of (3.1.7) and (3.1.8) may be stablished in a similar way for the formal case. Moreover, let $X^\wedge = \mathrm{Spec}\ (\hat{\mathcal{O}}_{X,P})$ where P is a closed point of X and let us denote by "$^\wedge$" the corresponding objects in $X^\wedge$ by the morphism $X^\wedge \longrightarrow X$. Then from (3.1.3.3) one deduces that

(3.1.9.1) $$\mathrm{Sing}^r(\mathcal{D}^\wedge_P, E^\wedge) = \mathrm{Sing}^r(\mathcal{D},E)^\wedge.$$

(3.2.) The directrix

(3.2.1) The aim of this paragraph is to estimate necessary conditions for the equality in (3.1.4.1) and (3.1.5.1). For this we shall use the concept of directrix or strict tangent space introduced by Hironaka (|10|, |11|).

(3.2.2) Let (R,η) be a local regular ring having k as a coefficient field and let $a \in R$ be such that $\nu_\eta(a) \geq 1$. Let H be the minimum k-vector subspace of $\mathrm{Gr}^1_\eta(R)$ such that

(3.2.2.1) $$\mathrm{In}\ (a) \in k[H] \subset \mathrm{Gr}_\eta(R).$$

Let $J(a) = H.\mathrm{Gr}_\eta(R)$, the "directrix of a" is the subscheme $V(J(a)) \subset \mathrm{Spec}\ (\mathrm{Gr}_\eta(R))$. For an $r \geq \nu(a)$, we shall denote $J^r(a) = J(a)$, if $r = \nu(a)$ and $J^r(a) = 0$, if $r > \nu(a)$.

(3.2.3.) Lemma. With notations as above, let $Z = \text{Spec}(R)$ and $\pi: Z' \longrightarrow Z$ the blowing-up of Z with center on its closed point P. Let $Y = V(a)$ and let Y' be the strict transform of Y by π. Let P' be a closed point of Z' such that $\pi(P') = P$. Assume that Y' is given at P' by $a' \in \mathcal{O}_{X'P'}$ and that $\nu(a') = \nu(a)$, then

(3.2.3.1) $$P' \in \text{Proj}\,(V(J(a))) \subset \pi^{-1}(P).$$

Proof. (See |10|).

(3.2.4) Let $\mathcal{D}$ be a multiplicatively irreducible unidimensional distribution (for short, m.i.u.d.) and adapted to E and let P be a closed point of X. Let $x_1,\dots,x_n$ be a r.s. of p. of $\mathcal{D}_{X,P}$ suited for (E,P) such that E is given by $\prod_{i\in A} x_i$. Let $\mathcal{D}_P$ be generated by

(3.2.4.1) $$D = \sum_{i\in A} a_i x_i \partial/\partial x_i + \sum_{i\notin A} a_i \partial/\partial x_i .$$

Finally, let us suppose that $r = \nu(\mathcal{D},E,P) \geq 1$. Now, in view of the equations (2.2.5.10), (2.2.5.11) and (2.2.5.12) (see also the proof of (3.1.4)) one has that (if $\mu = \mu(\mathcal{D},E,P)$) : $\mu = r$ iff there is no $i \notin A$ such that $\nu(a_i) = r$ and $\mu = r-1$ otherwise

(3.2.5) Definition. With notations as above let us define

(3.2.5.1)
$$J(\mathcal{D},E,P) = \sum_{i\in A} J^r(a_i) \quad \text{if } \mu = r.$$
$$J(\mathcal{D},E,P) = \sum_{i\notin A} J^r(a_i) \quad \text{if } \mu = r-1.$$

The "directrix of $\mathcal{D}$ at P" is the subscheme $V(J(\mathcal{D},E,P)) \subset \text{Spec}\,(\text{Gr}(\mathcal{O}_{X,P}))$ and it is denoted by $\text{Dir}\,(\mathcal{D},E,P)$.

(3.2.6) Remark. The ideal $J(\mathcal{D},E,P)$ of $\text{Gr}(\mathcal{O}_{X,P})$ does not depend on the choice of the generator D of $\mathcal{D}_P$ nor on the choice of the r.s. of p. suited for (E,P).

(3.2.7) Proposition. Let $X,\mathcal{D},E,P$ be as in (3.2.4), let $\pi: X' \longrightarrow X$ be the blowing-up of X with center P, let P' be a closed point of X' with $\pi(P') = P$ and let us suppose that $\nu(\mathcal{D},E,P) = \nu((\alpha(\mathcal{D}^\pi),E'),E',P')$ then

(3.2.7.1) $$P' \in \mathrm{Proj}\,(\mathrm{Dir}\,(\mathcal{D},E,P)) \subset \pi^{-1}(P).$$

Proof. Let us adopt the notations of 3.2.4. Let $\mu = r-1$, assume that a_i, $i \notin A$, is such that $\nu(a_i) = r$ and that a'_i is the strict transform of a_i by π (i. e. , by the blowing-up induced in $\mathrm{Spec}(\mathcal{O}_{X,P})$) then, from the equations of (2.2.5.10), (2.2.5.11) and (2.2.5.12) we have that $\nu(a'_i) = r$. If $\mu = r$, the same argument works for the a_i, $i \in A$, Now, the result follows from the lemma (3.2.3).

(3.2.8) (Remark that we only define the directrix at a closed point). In the formal case the directrix may be defined exactly in the same way as in (3.2.5). Also the result (3.2.7) remains true.

(3.2.9) Remark. The behaviour of the directrix of (3.2.5) is not as good as in the case of the analogous concept for varieties. As we shall see in (3.4.11), in the case of monoidal blowing-ups one has not a result as clean as (3.2.7). Moreover, even in the case of quadratic blowing -up, the dimension of the directrix may be increased by the blowing-up. For instance, if $\mathcal{D}_P$ is generated by

(3.2.9.1) $$D = (zy).x\partial/\partial x + z^3\partial/\partial y + x^5\partial/\partial z$$

where E is given by $x = 0$. Here $J(\mathcal{D},E,P) = (\underline{y},\underline{z})$ and the dimension of the directrix is one. Let us make the quadratic blowing-up in the direction $\underline{y} = \underline{z} = 0$ indicated by the directrix. The strict transform is generated at this point by

(3.2.9.2) $$D' = (x'y')x'\partial/\partial x' + (z'^3 - y'^2z')\partial/\partial y' + (x'^2-y'x'^2)\partial/\partial z'.$$

The adapted order is the same one but now the directrix is given by $\underline{x'} = 0$ and it has dimension two.

(3.3) Stationary sequences

(3.3.1) Unless $n = 2$, it is not possible in general to reduce the adapted order by making successive quadratic blowing-ups. (in |3|,|8|,|13| the proof for n = 2 is made). This paragraph is devoted to identify those curves which generate stationary situations when one makes blowing -up along their sequence of infinitely near points.

(3.3.2.) Let Y be a regular curve of X having normal crossings with E and let $P \in Y$ be a closed point. Let $\mathcal{D}$ be a m.i.u.d. over X and adapted to E. The sequence

(3.3.2.1) $$(\pi(t), X(t), E(t), Y(t), P(t), \mathcal{D}(t))$$

t = 0,1,..., is obtained inductively as follows:

a) X(o) = X, E(o) = E. Y(o) = Y. P(o) = P. $\mathcal{D}(o) = \mathcal{D}$.

b) $\pi(t)$: $X(t) \longrightarrow X(t-1)$ is the blowing-up of X(t-1) with center in P(t-1).

c) Y(t) is the strict transform of Y(t-1) by $\pi(t)$.

d) P(t) is the only closed point in $\pi(t)^{-1}(P(t-1)) \cap Y(t)$.

e) $E(t) = \pi(t)^{-1}(E(t-1) \cup P(t-1))$ with its reduced structure.

f) $\mathcal{D}(t)$ is the strict transform of $\mathcal{D}(t-1)$ by $\pi(t)$ adapted to E(t-1).

Let us observe that for $t \geq 1$ one has that for a component F of E(t)

(3.3.2.2) $$Y(t) \not\subset F$$

For the sake of simplicity let us assume that (3.3.2.2) is also true for t = 0. We shall denote

(3.3.2.3) $$r(t) = \nu(\mathcal{D}(t), E(t), P(t)).$$

(3.3.2.4) $$\mu(t) = \mu(\mathcal{D}(t), E(t), P(t)).$$

(3.3.3) <u>Definition</u>. The sequence (3.3.2.1) is "stationary" iff r(t)=r(0) for all t.

(3.3.4) Let $x = (x_1,\dots,x_n)$ be a r.s. of p. suited for (E,Y) at P. Then $\# B = n-1$ and if $i_o \notin B$ one has that $i_o \in A$ by (3.3.2.2). (see (2.2.1) for notations on A and B). Any $a \in \mathcal{O}_{X,P}$ may be expressed in exactly one way as $a = \sum a_I x^I$ where $a_I \in k$, $I \in \mathbb{N}^n$ and if $I=(i_1,\dots,i_n)$, $x^I = x_1^{i_1}\dots x_n^{i_n}$. Let us denote

(3.3.4.1) $$\mathrm{Exp}\,(a,x) = \{I;\ a_I \neq 0\} \subset \mathbb{Z}^n.$$

Now, let us assume that $r = r(0) \geq 1$, $\mu = \mu(0)$ and that $\mathcal{D}_P$ is generated by

(3.3.4.2) $$D = \sum_{i \in A} a_i x_i \partial/\partial x_i + \sum_{i \notin A} a_i \partial/\partial x_i .$$

Let $A = A^o$, $\{1,\dots,n\} - A = A^1$. Let

(3.3.4.3) $$\gamma^l(\mathcal{D},E,Y,P,x) = \min\{ h_{i_o} / (\mu+l- \sum_{j\in B} h_j);$$

$$\underline{h} = (h_1,\dots,h_n) \in \mathrm{Exp}\,(a_i,x),\ i \in A^l,\ \mu+l - \sum_{j\in B} h_j > 0\}$$

for $l = 0,1$ and let us denote

(3.3.4.4) $$\gamma(\mathcal{D},E,Y,P,x) = \min\,(\gamma^o,\gamma^1)$$

where $\gamma^l = \gamma^l(\mathcal{D},E,Y,P,x)$, $l = 0,1$. Here we assume $\min\,(\emptyset) = \infty$. Finally, and only for the case $\mu = r$, let us denote

(3.3.4.5)
$$\delta^l = \min\{ 1-l+ \sum_{j\in B} h_j;\ \underline{h} \in \mathrm{Exp}\,(a_i,x),\ i \in A^l\}\quad l=0,1$$
$$\delta(\mathcal{D},E,Y,P,x) = \min\,(\delta^o, \delta^1).$$

(3.3.5) The invariants γ and δ are actually independent of the choice of the suited x. Let us denote $x(0) = x$ and let $x(t)$ be obtained from

(3.3.5.1) $$x_{i_o}(t) = x_{i_o}(t-1);\ x_i(t)x_{i_o}(t) = x_i(t-1)\quad i \in B,$$

then $x(t)$ is a r.s. of p. of $\mathcal{O}_{X(t),P(t)}$ which is suited for the pair $(E(t),Y(t))$. Let us denote for short

(3.3.5.2) $$\gamma(t) = \gamma(\mathcal{D}(t),E(t),Y(t),P(t),x(t))$$

etc.

(3.3.6) <u>Lemma</u>. Assume that $\mu(0) = r(0)-1$. Then

a) If $\gamma(0) > 2$, then $r(1) = r(0)$, $\mu(1) = r(1)-1$ and $\gamma(1) = \gamma(0)-1$.

b) If $\gamma(0) < 2$, then $r(1) < r(0)$.

c) If $\gamma(0) = 2$, then $r(1) < r(0)$ or $r(1) = r(0)$, $\mu(1) = r(1)-1$ and $\gamma(1) = 1$.

Proof. Looking at (3.3.5.1), the result follows by looking at the monomials in the coefficients in view of the equations (2.2.5.10), (2.2.5.11) and (2.2.5.12).

(3.3.7) Lemma. Assume that $\mu(0) = r(0)$. Then

a) If $\gamma(0) \geq 2$, then $r(1) = r(0)$, $\mu(1) = r(1)$, $\delta(1) = \delta(0)$, $\gamma(1) = \gamma(0)-1$ and $\gamma^1(0) < \gamma^0(0)$ iff $\gamma^1(1) < \gamma^0(1)$.

b) If $\gamma(0) < 2$, $\gamma^1(0) < \gamma^0(0)$ and $\delta(0) = r$ then $r(1) = r(0)$, $\mu(1) = r(1)-1$ and $\gamma(1) = \infty$.

c) If $\gamma(0) < 2$ and $\delta(0) < r$ or $\gamma^0(0) \leq \gamma^1(0)$, then $r(1) < r(0)$.

(Note that in b) the only possibility is $\gamma(0) = 1$).

Proof. Similar to the proof of (3.3.6).

(3.3.8) Proposition. The sequence (3.3.2.1) is stationary iff one of the following statements is verified:

a) $\mu(0) = r(0)-1$ and $\gamma(0) = \infty$.

b) $\mu(0) = r(0)$ and $\gamma(0) = \infty$.

c) $\mu(0) = r(0)$, $\gamma^1(0) < \gamma^0(0)$ and $\delta(0) = r$.

Moreover, if c) holds, at the step $t = \gamma(0)$ one has that $\mu(t) = r(t)-1$ and $\gamma(t) = \infty$.

Proof. If follows from (3.3.6) and (3.3.7).

(3.3.9) The sequence (3.3.2.1) is stationary iff one has residually the situations a) or b) of (3.3.8). For an $a \in \mathcal{O}_{X,P}$, let us denote by $\nu_Y(a)$ the order of a with respect to the ideal of Y, then 3.3.8. a) is equivalent to

(3.3.9.1) "$\nu_Y(a_i) \geq r-1$ if $i \in A$, $\nu_Y(a_i) \geq r$ if $i \notin A$ and there exists $i \notin A$ such that $\nu(a_i) = r$".

where $\nu(a_i)$ denotes the order with respect to the maximal ideal of $\mathcal{O}_{X,P}$. Now 3.3.8. b) is equivalent to

(3.3.9.2) $$\text{"}\nu_Y(a_i) \geq r \text{ if } i \in A.\ \nu_Y(a_i) \geq r+1 \text{ if } i \notin A\text{"}.$$

(3.3.10) Remark. One can try to modify the concept of stationary sequence by requiring that $\mu(t) = \mu(0)$ for all t. But μ does not play in general the role of an adapted order. For instance, if $\mathcal{D}_P$ is generated by

(3.3.10.1) $$D = y^2 \cdot \frac{\partial}{\partial x} + y^3 \cdot y \frac{\partial}{\partial y} + z^3 \frac{\partial}{\partial z}$$

where E is given by $y = 0$, one has $\nu(\mathcal{D},E,P) = 2$ and $\mu(\mathcal{D},E,P) = 1$. If we make the blowing-up given by $x = x'$, $y = x'y'$, $z = x'z'$, then the strict transform is generated at P' by

(3.3.10.2) $$D' = y'^2 x' \frac{\partial}{\partial x'} + (x'y'^3 - y'^2)y' \frac{\partial}{\partial y'} + (x'z'^3 - y'^2 z') \frac{\partial}{\partial z'}.$$

The adapted order remains the same, but $\mu(\mathcal{D}',E',P') = 2$.

3.4. Permissible Centers

(3.4.1) Definition. Let $\mathcal{D}$ be a m.i.u.d. over X and adapted to E and let Y be a closed subscheme of X. We shall say that Y is a "weakly permissible center for $\mathcal{D}$ and adapted to E" iff the following conditions are verified:

a) Y is regular and it has normal crossings with E.

b) If $\mathcal{D}'$ is the strict transform of $\mathcal{D}$ adapted to E by the blowing-up $\pi: X' \longrightarrow X$ of X with center Y, then for any closed point P' of X' one has that

(3.4.1.1) $$\nu(\mathcal{D}',E',P') \leq \nu(\mathcal{D},E,\pi(P'))$$

(3.4.2) One can establish the same definition for the formal case changing the condition b) by

b)^ If $\hat{\pi}: \hat{X}' \longrightarrow \hat{X}$ is any directional blowing-up of $\hat{X}$ with center $\hat{Y}$

(see (2.3.2)) and if $\mathcal{D}^{\wedge}{}'$ is the strict transform of $\mathcal{D}^{\wedge}$, then

(3.4.2.1) $$\nu(\mathcal{D}^{\wedge}{}',E^{\wedge}{}',P^{\wedge}{}') \leq \nu(\mathcal{D}^{\wedge},E^{\wedge},P^{\wedge})$$

where $P^{\wedge}$ and $P^{\wedge}{}'$ are the closed points of $X^{\wedge}$ and $X^{\wedge}{}'$.

(3.4.3) The existence of stationary sequences as in (3.3) must a posteriori indicate the existence of a center for the blowing-up that one wants to use. The following definition is a sligth generalization of the conditions (3.3.9.1) and (3.3.9.2)

(3.4.4) Definition. Let $\mathcal{D}$,E,Y be as in (3.4.3) and let P be a closed point of Y. We shall say that Y is "permissible for $\mathcal{D}$ at P adapted to E" iff the following conditions are satisfied:

a) Y is regular and it has normal crossings with E at the point P.

b) Let $x = (x_1,\ldots,x_n)$ be a r.s. of p. of $\mathcal{O}_{X,P}$ suited for the pair (E,Y) and let us suppose that $\mathcal{D}_P$ is generated by

(3.4.4.1) $$D = \sum_{i\in A} a_i x_i \partial/\partial x_i + \sum_{i\notin A} a_i \partial/\partial x_i .$$

Then, one of the two following conditions is satisfied:

(3.4.4.2) "$\nu_Y(a_i) \geq r-1$ if $i \in A$ or $i \notin B$. $\nu_Y(a_i) \geq r$ if $i \in B-A$ and there exist $i \in B-A$ such that $\nu(a_i) = r$".

(3.4.4.3) "$\nu_Y(a_i) \geq r$ if $i \in A$ or $i \notin B$. $\nu_Y(a_i) \geq r+1$ if $i \in B-A$"

where $r = \nu(\mathcal{D},E,P)$.

We shall say that Y is "permissible for $\mathcal{D}$ adapted to E" iff it is so in all the closed points. Finally one can establish analogously the definition for the formal case.

(3.4.5) Remark. The conditions (3.4.4.2) and (3.4.4.3) are intrinsic in the sense that they do not depend on the suited r.s. of p. x.

On the other hand, the set of closed points of Y for which Y is permissible is open in the Zariski topology. Note that the conditions (3.3.9.1) and (3.3.9.2) are not open in P, because of the fact that E must have a transversal component to Y.

(3.4.6) Proposition. With notation as above. If Y is permissible for $\mathcal{D}$ adapted to E then Y is weakly permissible.

Proof. It is enough to prove (3.4.1) b). Let P be any closed point of Y. For P one has (3.4.4.2) iff

(3.4.6.1) $$\mu(\mathcal{D},E,Y) = \nu(\mathcal{D},E,P)-1$$

and one has (3.4.4.3) iff

(3.4.6.2) $$\mu(\mathcal{D},E,Y) = \nu(\mathcal{D},E,P).$$

Now the result follows in view of the equations of (2.2.5).

(3.4.7) Remark. Although $\mu(\mathcal{D},E,Y)$ does not depend on P, one can have (3.4.6.1) for a fixed P and (3.4.6.2) for another closed point P_1 of Y. In this case, necessarily

(3.4.7.1) $$\nu(\mathcal{D},E,P) = \nu(\mathcal{D},E,P_1) + 1.$$

For instance if $X = \mathbb{A}^3(k)$ and $\mathcal{D}$ is globally generated by

(3.4.7.1) $$D = (xy^2)x\partial/\partial x + z^3.y\partial/\partial y + z^3\partial/\partial z$$

where E is given by xy = 0. Let Y be y = z = 0. One obtains the example for the points $P = (0,0,0)$, $P_1 = (1,0,0)$.

(3.4.8) The same kind of reasonning proves the result (3.4.6) for the formal case.

(3.4.9) Proposition. With notations as in (3.4.4), let us suppose that Y is permisible for $\mathcal{D}$ and adapted to E at the closed point P and let us suppose that Y is a curve. Then, there is a finite sequence

(3.4.9.1) $(\pi(t), X(t), E(t), Y(t), Q(t), \mathcal{D}(t))$

$t = 0,1,\ldots,N$, such that

a) $X(0) = X$, $E(0) = E$, $Y(0) = Y$, $\mathcal{D}(0) = \mathcal{D}$ and $Q(0)$ is a closed point of $Y(0)$.

b) $\pi(t): X(t) \longrightarrow X(t-1)$ is the blowing-up with center $Q(t-1)$.

c) $E(t) = \pi(t)^{-1}(E(t-1) \cup \{Q(t)\})$.

d) $Y(t)$ is the strict transform of $Y(t-1)$ by $\pi(t)$.

e) $Q(t)$ is a closed point of $Y(t)$.

f) $\mathcal{D}(t)$ is the strict transform of $\mathcal{D}(t-1)$ by $\pi(t)$ adapted to $E(t-1)$.

g) Finally $Y(N)$ is permissible for $\mathcal{D}(N)$ and adapted to $E(N)$.

(i.e. the permissible centers can be "globalized").

Proof. First we can obtain that $Y(t)$ and $E(t)$ would have normal crossing by the general results of desingularization of curves (|1|, |10|,...). Now, in view of (3.4.5) there are only a finite number of points to consider. But each one will produce a stationary sequence and then the result follows from (3.3.8).

(3.4.10) The directrix does not work very well for monoidal blowing -up with permissible center. For instance, if $\mathcal{D}_P$ is generated by

(3.4.10.1) $$D = (xz^{r-1})\frac{\partial}{\partial x} + (xz^{r-1} + y^{2r})y\frac{\partial}{\partial y} + z^r \cdot \frac{\partial}{\partial z}$$

where E is given by $y = 0$. Here the directrix is given by $\underline{x} = \underline{z} = 0$ and it has dimension one. Let Y be given by $y = z = 0$, which is permissible at P. Then two things go "wrong":

a) The tangent space of Y is not contained in the directrix.

b) If we make the blowing-up $x = x'$, $y = y'$, $z = z'y'$, then the adapted order remains the same although the dimension of the directrix and of the center agree.

An example such that the tangent space of Y is contained in the directrix but b) holds may be generated by making

(3.4.10.2) $$D = (yz^{r-1})\partial/\partial x + (xz^{r-1} + y^{2r})y\frac{\partial}{\partial y} + z^r \cdot \frac{\partial}{\partial z}$$

and Y being as above.

(3.4.11) Proposition. With notations as in (3.4.4) assume that one of the two following statements holds:

a) dim Dir $(\mathcal{D},E,P) = n-1$.

b) $A \cup B = \{1,2,\ldots,n\}$.

Then one has that

i) The tangent space of Y at P is contained in Dir $(\mathcal{D},E,P)$.

ii) If $\pi: X' \to X$ is the blowing-up with center Y, P' is a closed point of X' with $\pi(P') = P$, $\mathcal{D}'$ is the strict transform of $\mathcal{D}$ adpated to E and $\nu(\mathcal{D}',E'P') = \nu(\mathcal{D},E,P)$ then one has that

(3.4.11.1) $$P' \in \mathrm{Proj}\ (\mathrm{Dir}\ (\mathcal{D},E,P)\ /\ T_PY) \subset \pi^{-1}(P)$$

where $T_P Y$ is the tangent space of Y at P.

Proof. Let us suppose that (3.4.4.3) holds. Then, in view of the equations of (2.2.5) since $\nu(\mathcal{D},E,P) = \nu(\mathcal{D}',E',P')$, then the strict transform of all a_i such that $\nu(a_i) = r$ and $i \in B-A$ must have order equal to $r = \nu(\mathcal{D},E,P)$. Since there exist at least one of this a_i, if a) holds, one has that

(3.4.11.2) $$J^r(a_i) = J(\mathcal{D},E,P)$$

If b) holds, one has that

(3.4.11.3) $$J(\mathcal{D},E,P) = \sum_{i\in B-A} J^r(a_i) = \sum_{i\in B-A,\ \nu(a_i)=r} J^r(a_i)$$

And then the result follows from the fact that

(3.4.11.4) $$P' \in \mathrm{Proj}\ (\sum_{i\in B-A} V(J^r(a_i))/T_PY).$$

If (3.4.4.4) holds we can apply the same kind of reasonement, but in this case we do not use the assumptions a) or b).

(3.4.12) As a consequence of the last remark in the preceeding proof, the statements i) and ii) of (3.4.11) always hold for a permissible center Y in the case that $\mu(\mathcal{D},E,Y) = \nu(\mathcal{D},E,P)$.

(3.4.13) Remark. (3.4.10), (3.4.11) and (3.4.12) may be also formulated for the formal case.

4. RESOLUTION STATEMENTS

(4.1.) The general statement

(4.1.1) Before making any statement, let us consider a few examples for illustrate some possible pathologies in the behaviour of the adapted order.

(4.1.2) Let $X = \mathbb{A}^2(k)$ and let $\mathcal{D}$ be generated globally by the vector field

(4.1.2.1) $$D = x.x \frac{\partial}{\partial x} + (y-m.yx) \frac{\partial}{\partial y}$$

where E is given by $x = 0$. Then the only singular point of $\mathcal{D}$ is the origin. Moreover, if we blow-up X with center the origin, there is only one closed point which is singular. The strict transform at this point is generated by

(4.1.2.2) $$D' = x'.x' \frac{\partial}{\partial x'} + (y'-(m+1)y'x') \frac{\partial}{\partial y'}$$

and we have a situation like the initial one. So $\mathcal{D}$ cannot be desingularized by quadratic blowing-ups.

(4.1.3) If Y is a divisor of X, then the blowing-up $\pi: X' \longrightarrow X$ of X with center Y is the identity, but if $\mathcal{D}$ is a m.i.u.d. over X adapted to E and Y has normal crossings with E, then the strict transform $\mathcal{D}'$ of $\mathcal{D}$ by π adapted to E is not exactly the same thing as $\mathcal{D}$. The difference is that $\mathcal{D}'$ must be considered as adapted to

(4.1.3.1) $$E' = \pi^{-1}(E \cup Y).$$

For instance, in the preceeding example if Y is given by $y = 0$, the strict transform is generated by

(4.1.3.2) $$D' = x'.x' \frac{\partial}{\partial x'} + (-mx')y' \frac{\partial}{\partial y'}$$

and it has no singular points.

(4.1.4) In the case of $n = 2$ it is not possible in general to reach adapted order equal to zero (see |3|) even if one admits transformations as in (4.1.3). For instance, let $X = \mathbb{A}^2(k)$ with $char(k) = 0$ and let $\mathcal{D}$ be globally generated by

(4.1.4.1) $$D = x.x \frac{\partial}{\partial x} + (y-(m!)x-myx) \frac{\partial}{\partial y}$$

where E is given by $x = 0$. Then $\mathcal{D}$ has only one singular point P = origin. If we make the blowing-up given by $x' = x$; $(y'+m!)x' = y$, then the strict transform is generated at this point P' by

(4.1.4.2) $$D' = x'.x' \frac{\partial}{\partial x'} + (y-(m+1)!.x-(m+1)yx) \frac{\partial}{\partial y}$$

which has the same form as D. Let us consider the sequence

(4.1.4.3) $$(\pi(t), X(t), E(t), P(t), \mathcal{D}(t))$$

given by blowing-up succesively the only singular point $P(t)$ of $\mathcal{D}(t)$. The sequence of $P(t)$ corresponds to the sequence of infinitely near points of the algebroid curve

(4.1.4.4) $$y - \sum_{i \geq m} i!.x^{i-m+1} = 0$$

which does not represent any curve in X. Now, let us suppose that in any step (all steps are similar, so we shall suppose $t = 0$) one makes the blowing-up with center a curve Y in such a way that we shall obtain $\mathcal{D}'$ with all points being regular. Now let us denote by $\mathcal{D}'(t)$ the succesive strict transforms of $\mathcal{D}'$ by $\pi(t)$. In view of (4.1.4.4), there is a step t_o such that $P(t_o)$ does not belong to $Y(t_o)$ (succesive transform of Y) and thus one has

(4.1.4.5) $$\mathcal{D}(t_o)_{P(t_o)} = \mathcal{D}'(t_o)_{P(t_o)}$$

which is a contradiction, because $P(t_o)$ must be a regular point of $\mathcal{D}'(t_o)$ while it is a singular point of $\mathcal{D}(t_o)$.

(4.1.5) First statement. Let $\mathcal{D}$ be a multiplicatively irreducible unidimensional distribution over X and adapted to a normal-crossings divisor E. Then there is a finite sequence

$$(4.1.5.1) \qquad (\pi(t), X(t), E(t), Y(t), \mathcal{D}(t))$$

$t = 0,1,\ldots,N$, such that

a) $X(0) = X$, $E(0) = E$, $\mathcal{D}(0) = \mathcal{D}$ and $Y(0)$ is a weakly permissible center for $\mathcal{D}$ adapted to E with $\dim Y(0) \leq n-2$.

b) $\pi(t)\colon X(t) \longrightarrow X(t-1)$ is the blowing-up of $X(t-1)$ with center $Y(t-1)$.

c) $E(t) = \pi(t)^{-1}(E(t-1) \cup Y(t-1))$.

d) $\mathcal{D}(t)$ is the strict transform of $\mathcal{D}$ by $\pi(t)$ adapted to $E(t-1)$.

e) $Y(t)$ is a weakly permissible center for $\mathcal{D}$ adapted to $E(t)$, such that $\dim Y(t) \leq n-2$.

f) Finally, for each closed point P of X(N) one has that

$$(4.1.5.2) \qquad \nu(\mathcal{D}(N), E(N), P) \leq 1.$$

(4.1.6) Remark. A strong version of (4.1.5) may be obtained by substituting "weakly permissible" by "permissible". Both versions are true in the case $n = 2$ for an arbitrary characteristic (| 3 |, | 8 |). The result (4.1.5) remains conjectural for $n \geq 3$. Moreover (at least in view of the techniques in this work) the result (4.1.5) for a fixed n seems to be strongly dependent on the desingularization results for hypersurfaces in dimension n-1. So the way in positive characteristic in hard. On the other hand, for low n, results as (4.1.5) are useful for desingularization of schemes in dimension n (| 7 |, | 9 |).

(4.1.6) Second statement. The sequence (4.1.5.1) may be chosen in such a way that for each closed point P of X(N), the following statement is verified: Let X^ be the scheme Spec $(\mathcal{O}_{X(N),P})$ and let $\mathcal{D}^\wedge, E^\wedge$ be the objects corresponding to $\mathcal{D}(N)$ and $E(N)$. Then, there is a normal crossings divisor $D^{\wedge\prime}$ of X' such that

a) $E^{\wedge\prime} \supset E^\wedge$.

b) If $\mathcal{D}^{\wedge\prime} = (\mathcal{D}^\wedge, E^{\wedge\prime})$ then one has that

(4.1.6.1) $$\nu(\mathcal{D}^{\wedge\prime}, E^{\wedge\prime}, P) = 0$$

(Let us observe that $\mathcal{D}^{\wedge\prime}$ may be obtained by a sequence of adapted blowing--ups with center in hypersurfaces).

(4.2) Resolution games

(4.2.1) In this paragraph we shall enounce the main result in this work, which is a "punctual" version of (4.1.5) and it is formulated in terms of a game between two players.

(4.2.2) Let $X^\wedge = \mathrm{Spec}\,(R)$ where R is a complete local regular ring of dimension n having k as a coefficient field. Let P be the closed point of $X^\wedge$, let $E^\wedge$ be a n.c. divisor on $X^\wedge$ and finally let $\mathcal{D}^\wedge$ be a formal m.i.u.d. over $X^\wedge$. The "reduction game" between two players A and B is defined as follows:

(4.2.3) Definition. Let $r = \nu(\mathcal{D}^\wedge, E^\wedge, P)$ and assume that $r \geq 2$. Let us denote $X^\wedge(0) = X^\wedge$, $\mathcal{D}^\wedge(0) = \mathcal{D}^\wedge$, $E^\wedge(0) = E^\wedge$, $P(0) = P$. We shall define "status t" and "movement t+1" for $t = 0,1,\ldots,$ inductively as follows

a) status 0 $= (X^\wedge(0), E^\wedge(0), \mathcal{D}^\wedge(0), P(0))$.

b) If the status t is

(4.2.3.1) $$(X^\wedge(t), E^\wedge(t), \mathcal{D}^\wedge(t), P(t)),$$

then the "movement t+1" runs in the following way: first the player A chooses a weakly permissible center $Y^\wedge(t)$ for $\mathcal{D}^\wedge(t)$ adapted to $E^\wedge(t)$ such that $\dim Y^\wedge(t) \leq n-2$ Second the player B chooses a directional blowing-up

(4.2.3.2) $$\pi^\wedge(t+1): X^\wedge(t+1) \longrightarrow X^\wedge(t)$$

of $X^\wedge(t)$ with center $Y^\wedge(t)$.

c) Let $\mathcal{D}(t+1)$ be the strict transform of $\mathcal{D}^\wedge(t)$ by $\pi^\wedge(t+1)$ and adapted to $E^\wedge(t)$. Let $E^\wedge(t+1) = \pi^\wedge(t+1)^{-1}(E^\wedge(t) \cup Y^\wedge(t))$ and let $P(t+1)$ be the closed point of $X^\wedge(t+1)$. Now, the "status t+1" is the 4-upla

(4.2.3.3) $$(X^\wedge(t+1), E^\wedge(t+1), \mathcal{D}^\wedge(t+1), P(t+1)).$$

Finally, the player A "wins at the movement t" iff

(4.2.3.4) $$\nu(\mathcal{D}(t), E^\wedge(t), P(t)) < r,$$

in this case, the game must stop at the status t. The player B wins iff A does not win at any movement, i.e. the game becomes infinite.

(4.2.4) The movement t may be identified with the pair $(Y^\wedge(t), \pi^\wedge(t+1))$. For short

(4.2.4.1) $$\mathrm{mov}(t) = (Y^\wedge(t), \pi^\wedge(t+1)).$$
$$\mathrm{stat}(t) = (X^\wedge(t), E^\wedge(t), \mathcal{D}(t), P^\wedge(t)).$$

(4.2.5) Definition. A "realization of the reduction game" is a (finite or infinite) sequence

(4.2.5.1) $$G = \{G(t) = (\mathrm{mov}(t), \mathrm{stat}(t))\} \quad t=0,1,\ldots$$

which respects the rules of (4.2.3) and such that if the last element corresponds to $t = N$, then

(4.2.5.2) $$\nu(\mathcal{D}(N), E(N), P(N)) < r.$$

(4.2.6) Definition. A "winning strategy" for the player A is a sequence of functions $F(t)$, $t = 1,2,\ldots$ in such a way that

a) $F(t)$ is defined over the set of sequences (which will be called "partial realizations")

(4.2.6.1) $$G|_t = \{G(s)\}_{0 \leq s \leq t-1}$$

where G is a realization of the reduction game, such that G(t-1) exists.

b) F(t)(G) is a weakly permissible center of $\mathcal{D}^\frown(t)$ adapted to $E^\frown(t)$ (note that mov(t-1) always produces stat (t) and stat (t) is known from $G|_t$).

c) If G is a realization of the reduction game which verifies the property

(4.2.6.2) $$\mathrm{mov}(t) = (F(t)(G|_t), \pi^\frown(t+1))$$

for all t, then G is finite (so player A wins).

If we put the following condition d) instead of c), we shall say that F(t) , t=1,2,... is a "strongly winning strategy":

d) There is an integer N such that if G verifies the property (4.2.6.2), then G is finite of length smaller than N.

(4.2.7) Remarks 1. A strong version of the reduction game and the concepts of winning strategy and strongly winning strategy may be obtained by putting "permissible" instead of "weakly permissible".

2. One can modify the reduction game by putting

(4.2.7.1) $$\nu(\mathcal{D}(t), E(t), P(t)) \leq 1$$

instead of (4.2.3.4). One can see easily that if there is a winning or strongly winning strategy for the new game, then there is so for the old one and that if there is a winning strategy for the old one there is so for the new one. But if there is a strongly winning strategy for the reduction game, one needs to assure that once obtained the reduction of the adapted order as in (4.2.3.4), the new possible starting status generates only a finite number of integers N as in d) in order to obtain the existence of a strongly winning strategy for the new game.

(4.2.8) If (4.1.5) is true, then there is a strongly winning strategy for the reduction game each time that the status 0 comes from a situation over X. Actually, the strategy may be defined in such a way that all possible realizations may be obtained by choosing directional blowings-up of the $\pi(t)$ of (4.1.5.1).

(4.2.9) The main result in this work is the following.

Theorem. Let $n = 3$ and let car $k = 0$, then there exists a winning strategy for the reduction game.

A PARTIAL WINNING STRATEGY

0. INTRODUCTION

(0.0.1) In this chapter we shall begin the proof of the existence of a winning strategy for the reduction game in the case $n = 3$ and char $k = 0$. We shall restrict the possibilities of the status 0 of the resolution game to those called "of the type zero" in order to prove the existence of a winning strategy in this case. Let us remark that no assumption on the characteristic is made in any part of the chapter.

(0.0.2) Before starting, let us consider some general reductions for the possibilities of the status 0 of the reduction game. Let us suppose that

(0.0.2.1) $$\text{stat}\ (0) = (X(0), E(0), \mathcal{D}(0), P(0))$$

is the status 0 of the reduction game (in the sequel we shall not use the symbol "^" anymore, since all the distributions, etc. will be "formal"). And let us suppose that

(0.0.2.2) $$\dim \text{Dir}\ (\mathcal{D}(0), E(0)) = 0.$$

Then, by (I. 3.2.7) if the player A chooses the closed point P(0) as center, then he always wins in the first movement. Thus we shall suppose that (0.0.2.2) does not hold and the reduction game may be modified by saying that the player A wins at the movement t iff one has

(0.0.2.3) $$\nu(\mathcal{D}(t),E(t),P(t)) < r$$

$$\text{or} \quad \dim \operatorname{Dir}(\mathcal{D}(t),E(t)) = 0.$$

where $r = \nu(\mathcal{D}(0),E(0),P(0))$.

(0.0.3) On the other hand let us remark that one has always $E(t) \neq \emptyset$ for $t \geq 1$. Thus if one looks for a winning strategy, there is no loss of generality in assuming that

(0.0.3.1) $$E(0) \neq \emptyset.$$

In the sequel, we shall make this assumption. This reduction would not be so easy if we were looking for a strongly winning strategy. One would have to control the possible status 1 after the first move.

1. TYPE ZERO SITUATIONS

(1.1) Description of type zero

(1.1.1) Let $X = \mathrm{Spec}\ (R)$, where R is a complete regular local ring of dimension three having k as a coefficient field. Let E be a normal crossings divisor on X and let P be the closed point of X. In the sequel $\mathcal{D}$ will denote a formal multiplicatively irreducible unidimensional distribution over X and adapted to E such that

$$r = \nu(\mathcal{D},E,P) \geq 2, \tag{1.1.1.1}$$

with all the assumptions of (0.0.2), (0.0.3).

(1.1.2) For a vector subspace V of the tangent space T_PX, let us consider the set

$$H(V) = \{f \in R;\ \nu(f) = 1,\ T_P\ (f=0) \supset V\}. \tag{1.1.2.1}$$

We shall say that V and $\mathcal{D}$ are in an "adapted to E transversal position" iff

$$0 \neq \sum_{f \in H(V)} J^r(D(f)) = J_{H(V)} \tag{1.1.2.2}$$

(D is a generator of $\mathcal{D}$) and there exists a linear form $\underline{z} \in J_{H(V)}$ such that $\underline{z} = 0$ is transversal to the tangent space of some component of E.

Let us remark that if $\nu(D(f)) = r$, then $J^r(D(f)) \subset J(\mathcal{D},E)$. So if dim Dir $(\mathcal{D},E) = 2$, by (1.1.2.2) one has that

$$J_{H(V)} = J(\mathcal{D},E). \tag{1.1.2.3}$$

(1.1.3) Definition. We shall say that $(X,E,\mathcal{D},P)$ is of the "type zero" iff the following two conditions are verified

a) dim Dir $(\mathcal{D},E) \geq 1$ and $E \neq \emptyset$.

b) $\mathcal{D}$ and Dir $(\mathcal{D},E)$ are in an adapted to E transversal position.

(1.1.4) Lemma. $(X,E,\mathcal{D},P)$ is of type zero iff there is a r.s. of p. (x,y,z) suited for (E,P) such that

i) E is given by x or by xy.

ii) $(\underline{x},\underline{y},\underline{z}) \neq J(\mathcal{D},E) \supset (\underline{z})$.

iii) Let ∂_y be $\partial/\partial y$ or $y\,\partial/\partial y$ according to E. Then if

(1.1.4.1) $$D = ax\partial/\partial x + b\partial_y + c\partial/\partial z$$

generates $\mathcal{D}$, one has that $\nu(c) = r$.

iv) If dim Dir $(\mathcal{D},E) = 1$ and E has only one component, then one of the following properties is satisfied:

(1.1.4.2) $$J(\mathcal{D},E) = (\underline{y},\underline{z})$$

(1.1.4.3) $$J(\mathcal{D},E) = J^r(c) = (\underline{x},\underline{z})$$

(1.1.4.4) $$J(\mathcal{D},E) = (\underline{x},\underline{z});\ J^r(c) = (\underline{z}).$$

Proof. Let us suppose that $(X,E,\mathcal{D},P)$ is of the type zero. We can choose (x,y,z) such that E is given by x or xy because E cannot have three components (otherwise $\nu(D(f)) > r$ for all f). Since there is $f \in H(\mathrm{Dir}(\mathcal{D},E))$ with $\nu(D(f)) = r$, one has that $\alpha\underline{x} + \beta\underline{y} + \gamma\underline{z} \in J(\mathcal{D},E)$ where $(\beta,\gamma) \neq (0,0)$ if E has one component and $\gamma \neq 0$ if E has two components. By making an adecuate change of coordinates we may suppose $\underline{z} \in J(\mathcal{D},E)$ and so we have ii). If dim Dir $(\mathcal{D},E) = 2$, then $J(\mathcal{D},E) = (\underline{z})$ and iii) follows from the fact that $D(z) = c$. Let us suppose that dim Dir $(\mathcal{D},E) = 1$. If E has two components, necessarily $\nu(c) = r$ since otherwise $\nu(D(f)) > r$ for all f. If E has one component, let us distinguish the two following posibilities: $J(\mathcal{D},E)=(\underline{x},\underline{z})$ or $J(\mathcal{D},E) = (\underline{y} + \alpha\underline{x},\underline{z})$. In the second case, by an adecuate change of coordinates we have $\alpha = 0$ and if $\nu(c) > r$ we shall interchange y and z. In the first case if $J^r(c)=(\underline{x},\underline{z})$ there is no problem, if $\nu(c) > r$ then $\nu(D(f)) > r$ for all $f \in H(\mathrm{dir}(\mathcal{D},E))$. So we may assume that $J^r(c) = (\alpha\underline{x} + \beta\underline{z})$, if $\beta \neq 0$, a change of coordinates gives (1.1.4.4). Now, if $J^r(c) = (\underline{x})$ there is no linear form in

(1.1.4.5) $$\sum_{f \in H(\mathrm{Dir}(\mathcal{D},E))} J^r(D(f))$$

transversal to the tangent space of E.

Conversely, from i) and ii) we have (1.1.3) a). If dim $\mathrm{Dir}(\mathcal{D},E) = 2$ then $J(\mathcal{D},E) = (\underline{z})$ and (1.1.3) b) follows from iii). If dim Dir $(\mathcal{D},E) = 1$ and E has two components, then $J(\mathcal{D},E) = J^r(c) \supset (\underline{z})$ and (1.1.3) b) follows easily. If E has one component and $J(\mathcal{D},E) = (\underline{y},\underline{z})$ then $J^r(c) = (\underline{y},\underline{z})$ or $J^r(c) = (\alpha\underline{y} + \beta\underline{z})$ with $(\alpha,\beta) \neq$ $\neq (0,0)$, in both cases it is enough to consider $z \in H(Dir(\mathcal{D},E))$. Analogously if $J^r(c) = (\underline{x},\underline{z})$ or $J^r(c) = (\underline{z})$.

(1.1.5) Remark. The situation (1.1.4.4) is the only one for which one has that

(1.1.5.1) $$\sum_{f \in H(Dir(\mathcal{D},E))} J^r(f) \neq J(\mathcal{D},E)$$

so, in some sense, is a "less transversal situation" than the other's.

(1.1.6) Definition. $(X,E,\mathcal{D},P)$ is of the "type 0-1" iff it is of the type zero and dim $Dir(\mathcal{D},E) = 2$ or one has the property (1.1.5.1). Otherwise it is of the "type 0-0".

(1.2) Stability results

(1.2.1) Theorem. With notations as in (1.1.1). Let us suppose that $(X,E,\mathcal{D},P)$ is of the type 0-0 and that $(X',E',\mathcal{D}',P')$ is a directional quadratic blowing-up of $(X,E,\mathcal{D},P)$. Then one of the two following possibilities is satisfied :

a) $\nu(\mathcal{D}',E',P') < r$ or dim Dir $(\mathcal{D}',E') = 0$

b) $\nu(\mathcal{D}',E',P') = r$ and $(X',E',\mathcal{D}',P')$ is of the type 0-0.

Proof. Assume that a) is not satisfied and let (x,y,z) be as in (1.1.4). There are two possibilities: 1º (x,y,z) may be chosen in such a way that $J(\mathcal{D},E) =$ $= (\underline{y},\underline{z})$ besides the conditions of (1.1.4). 2º not 1º. In the first case we must consider the blowing-up given by

(1.2.1.1) $$x = x'; \ y = x'y'; \ z = x'z'$$

in view of (I. 3.2.7). If E has two components, then $c' = c/x'^r - z'a'$ (a',b' and c' coefficients of a generator of $\mathcal{D}'$) in view of (I. 2.2.5), so, since a) is not satisfied, one has that

(1.2.1.2) $$\mathrm{In}\,(c') = \mathrm{In}\,(c/x'^{r})$$

(the initial form being referred to the local ring of X' at P'), thus, since a) is not satisfied

(1.2.1.3) $$\dim \mathrm{Dir}\,(\mathcal{D}',E') = \dim \mathrm{Dir}\,(c' = 0) = 1$$

$$J^{r}(c') \supset (\underline{z}' + \lambda \underline{x}')$$

(see in |10| vgr. the behaviour of the directrix of an hypersurface). So a change $z'_1 = z' + \lambda x'$ leads us to the situation of lemma (1.1.4). Let us suppose that E has only one component. If $J^{r}(c) = (\underline{y},\underline{z})$, or simetrically $J^{r}(b) = (\underline{y},\underline{z})$, we shall reason as above. Let us suppose that $J^{r}(b) = (\alpha\underline{y} + \beta\underline{z})$, $J^{r}(c) = (\gamma\underline{y} + \delta\underline{z})$ with (α, β) and (γ, δ) independents. We have $b' = b/x'^{r} - y'a'$, $c' = c/x'^{r} - z'a'$. Since dim Dir $(b' = 0) \geq 1$, one has that

(1.2.1.4) $$J^{r}(b') = (\alpha\underline{y}' + \beta\underline{z}' + \lambda\underline{x}') \text{ or } J^{r}(b') = (\alpha\underline{y}' + \beta\underline{z}',\underline{x}')$$

and analogously with c':

(1.2.1.5) $$J^{r}(c') = (\gamma\underline{y}' + \delta\underline{z}' + \underline{x}') \text{ or } J^{r}(c') = (\gamma\underline{y}' + \delta\underline{z}',\underline{x}').$$

Since a) is not satisfied, the only possibilities are the first ones in both cases, now it is enough to make the change $y'_1 = \alpha y' + \beta z' + \lambda x'$, $z'_1 = \gamma y' + \delta z' + \mu x'$, in order to obtain the situation of the lemma (1.1.4).

If one cannot assume $J(\mathcal{D},E) = (\underline{y},\underline{z})$, there are only the two following possibilities:

(1.2.1.6) "E has two components and $J(\mathcal{D},E) = (\underline{y}+\zeta\underline{x},\underline{z})$ with $\zeta \neq 0$".

(1.2.1.7) "E has one component and $J^{r}(c) = (\underline{x},\underline{z})$".

In the case (1.2.1.6) the blowing-up is given by

(1.2.1.8) $$x = x';\ y = (y'-\zeta)x';\ z = x'z',$$

$c' = c/x'^{r} - z'a'$, dim Dir $(c'=0) = 1$ and

(1.2.1.9) $$J^r(c') = (y' + \lambda\underline{x}', \underline{z}' + \mu\underline{x}')$$

now it is enough to make the change $y'_1 = y' + \lambda x'$, $z'_1 = z' + \mu x'$, to obtain (1.1.4.2). In the case (1.2.1.7) the blowing-up is given by

(1.2.1.10) $$x = z'y';\ y = y';\ z = y'z'$$

and $c' = c/y'^r - z'b'$. As above, $\underline{z}' + \lambda\underline{y}' \in J^r(c')$ and dim Dir $(c'=0) = 1$. By making the change $z'_1 = z' + \lambda y'$, the proof is finished.

(1.2.2) Notation. We shall standarize following |*10*| the notation for the most frequently used equations of a blowing-up. We shall denote:

$$(T\text{-}1,\zeta): x = x';\ y = (y'-\zeta)x';\ z = x'z'.$$

$$(T\text{-}2) : x = x'y';\ y = y';\ z = y'z'.$$

$$(T\text{-}3) : x = x';\ y = y';\ z = x'z'.$$

$$(T\text{-}4) : x = x';\ y = y';\ z = y'z'.$$

(1.2.3.) Proposition. Assume that $(X,E,\mathcal{D},P)$ is of the type 0-0 and that Y is a permissible center which is tangent to the directrix Dir $(\mathcal{D},E)$. If $(X',E',\mathcal{D}',P')$ is a directional blowing-up of $(X,E,\mathcal{D},P)$ with center Y then one has that

(1.2.3.1) $$\nu(\mathcal{D}',E',P') < r.$$

Proof. The only case which does not correspond to a situation as in (I. 3.4.12) corresponds to the case (1.1.4.3). But in this case (x,y,z) may be taken in (1.1.4) with the property of being suited for (E,Y). Then the result follows from the fact that dim Dir $(c=0) = 1$.

(1.2.4) Proposition. Let us suppose that $(X,E,\mathcal{D},P)$ is of the type 0-1 and that $(X',E',\mathcal{D}',P')$ is a directional quadratic blowing-up of $(X,E,\mathcal{D},P)$. Then one of the following possibilities is satisfied

a) $\nu(\mathcal{D}',E',P') < r$ or dim Dir $(\mathcal{D}',E') = 0$.

b) $\nu(\mathcal{D}',E',P') = r$ and $(X',E',\mathcal{D}',P')$ is of the type 0-0.

c) $\nu(\mathcal{D}',E',P') = r$ and $(X',E',\mathcal{D}',P')$ is of the type 0-1.

Proof. Let us suppose that a) is not satisfied. Let (x,y,z) be as in lemma (1.1.4). In view of (I. 3.2.7), we must consider the equations $(T\text{-}1,\zeta)$ or T-2. We have that

(1.2.4.1) $$c' = c/x'^r - z'a' \quad \text{or} \quad c' = c/y'^r - z'b'$$

accordingly to $(T\text{-}1,\zeta)$ or T-2. Thus, if dim Dir $(c'=0) = 1$ one has necessarily $J^r(c') = (\underline{x}',\underline{z}')$ or $J^r(c') = (\underline{z}'+\lambda\underline{x}',\underline{y}' + \mu\,\underline{x}')$ and, after a change $z'_1 = z+\lambda x$ in the second case, we have type 0-0. If dim Dir $(c'=0) = 2$ one has that $J^r(c') = (\underline{z}'+\lambda\underline{x}')$ and then a change $z'_1 = z'+\lambda x'$ shows that we have type 0-1 if dim Dir $(\mathcal{D}',E') = 2$ or $J(\mathcal{D}',E') = (\underline{x}',\underline{z}')$ and type 0-0 otherwise.

(1.2.5) Remark. If $(X,E,\mathcal{D},P)$ is of the type 0-1 and (x,y,z) is like in (1.4.1) then a permissible center Y is always tangent to $\underline{z} = 0$. Actually, in the case dim Dir $(\mathcal{D},E) = 2$ there is nothing to do in view of (I. 3.4.2). In the case dim Dir $(\mathcal{D},E) = 1$, necessarily Y is given by (x,h) with $\underline{x}$ and In(h) independents, since otherwise we can apply (I. 3.4.2) too. We can suppose $\mathrm{In}(h) = \alpha\underline{y} + \beta\,\underline{z}$. If $\alpha \neq 0$ we can make $y_1 = h$ and then the permissibility is not possible since $\underline{y}_1$ does not divide In (b_1) (b_1 = coefficient of $\partial/\partial y_1$ in the base obtained from (x,y,z)). Thus $\alpha = 0$ and Y are tangent to $\underline{z} = 0$.

Let us observe that if we make $z_1 = h$ then (x,y,z_1) have the properties required in (1.1.4) and then Y is given by (x,z_1).

Moreover, an easy calculation shows that in this case if $(X',E',\mathcal{D}',P')$ is a directional blowing-up with center Y and $\nu(\mathcal{D}',E',P') = r$ then one has that

(1.2.5.1) $$P' \in \mathrm{Proj}[(\underline{z} = 0) \,/\, (\underline{x} = \underline{z} = 0)].$$

(1.2.6) Theorem. Let us suppose that $(X,E,\mathcal{D},P)$ is of the type 0-1, that Y is a permissible center and that $(X',E',\mathcal{D}',P')$ is a directional blowing-up of $(X,E,\mathcal{D},P)$ with center Y. Then one of the three following possibilities is satisfied:

a) $\nu(\mathcal{D}',E',P') < r$ or dim Dir $(\mathcal{D}',E') = 0$.

b) $\nu(\mathcal{D}',E',P') = r$ and $(X',E',\mathcal{D}',P')$ is of the type 0-0.

c) $\nu(\mathcal{D}',E',P') = r$ and $(X',E',\mathcal{D}',P')$ is of the type 0-1.

Proof. We shall suppose that a) is not satisfied. An easy calculation (see the above remark) shows that we can choose (x,y,z) as in (1.1.4) with the additional property that Y is given by (x,z) or by (y,z). In this situation (see (1.2.5.1)) we may suppose that the equations are T-3 or T-4. The rest of the proof is similar to the proof of (1.2.4).

(1.3) Type zero games

(1.3.1) The stability results of (1.2) allow a decomposition of the reduction game of (4.2.3) beginning at a status 0 of type zero into two reduction games corresponding to type 0-1 and type 0-0. Actually, we shall deal only with permissible centers (and not merely weakly permissible) and thus we are decomposing the strong version of the reduction game (see I. 4.2.7).

(1.3.2) Definition. The "reduction game of type 0-0" is defined as follows. Status 0 is a type 0-0 r-upla. Status t and movement t+1 are defined as in (I. 4.2.3) putting "permissible" instead of "weakly permissible". We shall say that the player A wins at the movement t iff one of the two following possibilities is satisfied

a) $\nu(\mathcal{D}(t),E(t),P(t)) < r$.

b) dim Dir $(\mathcal{D}(t),E(t)) = 0$.

(1.3.3) Definition. The "reduction game of type 0-1" is defined as follows. Status 0 is a type 0-1 4-upla. Status t and movement t+1 are defined as in (1.3.2). Player A wins at the movement t iff one of the three following possibilities is satisfied:

a) $\nu(\mathcal{D}(t),E(t),P(t)) < r$.

b) dim Dir $(\mathcal{D}(t),E(t)) = 0$.

c) $\nu(\mathcal{D}(t),E(t),P(t)) = r$ and the 4-upla $(X(t),E(t),\mathcal{D}(t),P(t))$ is of the type 0-0.

(1.3.4) Winning strategies and strongly winning strategies for both games are defined in the same way as in (I. 4.2.6). In this chapter we shall prove that one can found a winning strategy for each of the two games above. Thus, as a corollary of the stability results of (1.2) one deduces the existence of a winning strategy for the reduction game beginning at type zero.

2. A TYPE 0-0 WINNING STRATEGY

In this section we shall prove the existence of a winning strategy for the reduction game of type 0-0. It is quite simple: if at status t there is a permissible center tangent to the directrix, then player A chooses this center and wins, otherwise player A chooses the closed point as center. This strategy depends only on the status t and not on the history of the realization of the game.

The invariants used for the control of this game are of a kind similar to the one of the invariants used in |1| or |10| for the control of the singularities of curves or surfaces with one dimensional directrix.

(2.1) An invariant of transversality

(2.1.1) Let $(X,E,\mathcal{D},P)$ be of type 0-0. In view of (1.1.4) we have two possibilities:

a) Each linear form in

(2.1.1.1)
$$J = \sum_{f \in H(\mathrm{Dir}(\mathcal{D},E))} J^r(D(f))$$

is transversal to a fixed component of E.

b) For each component of E there is a linear form in J wich is tangent to it.

Moreover, in the case b) necessarily E has only one component (it corresponds to (1.1.4.3)).

(2.1.2) Definition. For a type 0-0 4-upla $(X,E,\mathcal{D},P)$ we shall define $w(\mathcal{D},E) = 1$ iff we have (2.1.1) b) and $w(\mathcal{D},E) = 0$ iff we have (2.1.1) a).

(2.1.2) Remark. Let us observe that $w = 1$ is "less transversal" than $w = 0$. In general, the evolution by blowing-ups goes from "weak" transversality to "strong" transversality.

(2.1.3) Lemma. If $w = 0$ the r.s. of p. (x,y,z) of (1.1.4) may be chosen in such a way that $J(\mathcal{D},E) \neq (\underline{x},\underline{z})$.

Proof. See the proof of (1.2.1).

(2.1.4) Definition. A r.s. of p. (x,y,z) as in (1.1.4) and with the additional property of (2.1.3) will be called a "normalized system of parameters for $(X,E,\mathcal{D},P)$".

(2.1.5) Proposition. Let $(X,E,\mathcal{D},P)$ be of the type 0-0 and let $(X',E',\mathcal{D}',P')$ be a directional quadratic blowing-up of $(X,E,\mathcal{D},P)$. Then one of the two following possibilities is satisfied:

a) $\nu(\mathcal{D}',E',P') < r$ or $\dim \mathrm{Dir}\,(\mathcal{D}',E') = 0$.

b) $w(\mathcal{D},E) \geq w(\mathcal{D}',E') = 0$.

Proof. If $w(\mathcal{D},E) = 1$, in view of (1.1.4.3) if a) is not satisfied, then for a normalized system of parameters we have to make T-2 (see 1.2.2) and then we are in the situations of the proof of (1.2.1) except for (1.2.1.7) and the result follows easily from that proof. If $w(\mathcal{D},E) = 0$, the proof is easy.

(2.1.6) Remark. We can suppose that $w = 0$, in order to find a winning strategy.

(2.2) Polygons for type 0-0

(2.2.1) In order to unify the techniques used, we shall introduce a polygon associated to a situation of the type 0-0 with $w = 0$ and a normalized system of parameters. But no full use of this idea will be made until the study of the type 0-1.

(2.2.2) Let p = (x,y,z) be a r.s. of p. of R, K the field of fractions of R and $f \in R[y^{-1}] \subset K$. Then f may be written in a unique way as

$$f = \sum f_{hij}\, x^h y^i z^j \tag{2.2.2.1}$$

where $f_{hij} \in k$. We shall define the "cloud of points of f with respect to p" by

$$\mathrm{Exp}\,(f,p) = \{(h,i,j);\ f_{hij} \neq 0\} \subset \mathbb{Z}^3. \tag{2.2.2.2}$$

We shall write Exp(f) if there is no confussion with p.

(2.2.3) Notation. Let us denote by e(E) the number of components of the normal crossings divisor E at P.

(2.2.4) Definition. Let $(X,E,\mathcal{D},P)$ be of the type 0-0 with $w(\mathcal{D},E) = 0$ and let p = = (x,y,z) be a normalized system of parameters for $(X,E,\mathcal{D},P)$. Assume that $\mathcal{D}$ is generated by

$$D = ax\partial/\partial x + b\partial_y + c\partial/\partial z \tag{2.2.4.1}$$

where $\partial_y = \partial/\partial y$ or $y\partial/\partial y$ accordingly to e(E). The "cloud of points Exp(D,E,p) of D adapted to E and with respect to p" is defined by:

$$\mathrm{Exp}(D,E,p) = \mathrm{Exp}(za) \cup \mathrm{Exp}(zb/y) \cup \mathrm{Exp}(c) \tag{2.2.4.2}$$

if e(E) = 1 and

$$\mathrm{Exp}(D,E,p) = \mathrm{Exp}(za) \cup \mathrm{Exp}(zb) \cup \mathrm{Exp}(c) \tag{2.2.4.3}$$

if e(E) = 2.

(2.2.5) Remark. Exp(D,E,p) depends actually on the generator D.

(2.2.6) Definition. With notations as in (2.2.4) the invariant $m(\mathcal{D},E,p)$ is defined by

(2.2.6.1) $$m(\mathcal{D},E,p) = \min\{h;\ (h,-1,r) \in Exp(D,E,p)\}$$

and $m(\mathcal{D},E,p) = +\infty$ if the set on the rigth is empty.

(2.2.7) Remark. $m(\mathcal{D},E,p)$ does not depend on the generator D.

(2.2.8) Let $\phi: \mathbb{R}^2 \to \mathbb{R}^3$ be the immersion given by $\phi(u,v) = (u,v,r-1)$ and let $\psi : \mathbb{R}^3 - \{(0,0,r)\} \to \mathbb{R}^2$ be the composition of ϕ^{-1} with the projection from $(0,0,r)$ onto $\phi(\mathbb{R}^2)$. Finally, given $m \in \mathbb{Z}_0 \cup \{\infty\}$, let $\mathbb{H}(m)$ be the set

(2.2.8.1) $$\mathbb{H}(m) = \{(u,v);\ u \geq 0,\ u+mv \geq 0\} \subset \mathbb{R}^2$$

$$\mathbb{H}(\infty) = \{(u,v);\ u \geq 0,\ v \geq 0\}$$

(2.2.9) Definition. With notations as above, "the polygon $\Delta(\mathcal{D},E,p)$ of $\mathcal{D}$ adapted to E and with respect to p" is defined by the convex hull of

(2.2.9.1) $$[\psi(Exp(\mathcal{D},E,p) \cap \{(h,i,j);\ j \leq r-1\}) + \mathbb{H}(m(\mathcal{D},E,p))] \cap \{(u,v);\ v \leq -1\}.$$

(2.2.10) Remark. $\Delta(\mathcal{D},E,p)$ does not depend on the generator of $\mathcal{D}$.

(2.2.11) Proposition. With notations as above, the curve Y given by (y,z) is permissible iff

(2.2.10.1) $$\Delta(\mathcal{D},E,p) \subset \{(u,v);\ v \geq 1\}$$

Proof. (y,z) is permissible iff $m(\mathcal{D},E,p) = \infty$ and for any $(h,i,j) \in Exp(D,E,p)$ such that $j \leq r-1$, one has that $\psi(h,i,j) = (h/(r-j),i/(r-j))$ satisfies that $i/(r-j) \geq 1$.

(2.2.12) Definition. With notations as above, the invariant $\delta(\mathcal{D},E,p)$ is defined by the lowest $t > 0$ such that

(2.2.12.1) $$\{(u,v),\ u+tv = t\} \cap \Delta(\mathcal{D},E,p)-\{(0,1)\} \neq \emptyset.$$

If there is no $t > 0$ satisfying (2.2.12.1) we put $\delta(\mathcal{D},E,p) = \infty$.

(2.2.13) Remarks a) $\delta(\mathcal{D},E,p) = \infty$ iff (y,z) is a permissible center.

b) $\delta(\mathcal{D},E,p) \geq 1$. This follows from the fact that $\nu(\mathcal{D},E,\mathcal{D}) = r$.

(2.2.14) Definition. With notations as above, we shall say that $p = (x,y,z)$ is "strongly normalized" iff

(2.2.14.1) $$(0,1) \in \Delta(\mathcal{D},E,p).$$

(2.2.15) Remark. There are normalized but not strongly normalized regular systems of parameters. For instance let D be

(2.2.15.1) $$D = x^r . x\partial/\partial x + (y+x)z^{r-1}\partial/\partial y + z^r . \partial/\partial z.$$

This is not "true" in the case of hypersurfaces: if $\nu(f) = r$, dim Dir $(f=0) = 1$, $\underline{z} \in J^r(f)$ and $J^r(f) \neq (\underline{x},\underline{z})$ then

(2.2.15.2) $$(0,1) \in \Delta(f,p)$$

where $\Delta(f,p)$ is the convex hull of $\psi(\mathrm{Exp}(f,p) \cap \{(h,i,j);\ j \leq r-1\}) + \mathbb{R}_0^2$.

(2.2.16) Proposition. If $(X,E,\mathcal{D},P)$ is of type 0-0 and $w(\mathcal{D},E) = 0$ there exists always a strongly normalized system of parameters.

Proof. If $e(E) = 2$, we deal only with the third coefficient of D and, in view of the above remark, "normalized" and "strongly normalized" are equivalent concepts. Let us suppose that $e(E) = 1$. Let $p = (x,y,z)$ be a normalized r.s. of p., in view of the definitions, the only possibility for $(0,1) \notin \Delta(\mathcal{D},E,p)$ is that

(2.2.16.1) $$\mathrm{In}(b) = \lambda\underline{y} \cdot \underline{z}^{r-1} + \mu\underline{z}^r + \underline{x}(\ldots)$$

with $\lambda \neq 0$. Then it is enough to interchange y and z.

(2.2.17) Proposition. If $(X,E,\mathcal{D},P)$ is of the type 0-0, $w(\mathcal{D},E) = 0$ and $p = (x,y,z)$ is strongly normalized, then $\delta(\mathcal{D},E,p) = 1$ iff $J(\mathcal{D},E) = (\underline{y}+\lambda\underline{x},\underline{z})$ with $\lambda \neq 0$.

Proof. If $e(E) = 2$ we deal only with c and the result follows from the standard facts about hypersurfaces (|10|). Let us suppose that $e(E) = 1$. Since $(0,1) \in \Delta(\mathcal{D},E,p)$ we have that $\delta(\mathcal{D},E,p) = 1$ iff $m(\mathcal{D},E,p) = 1$ or there is a point different from (0,1) in

(2.2.17.1) $$\psi(\mathrm{Exp}(D,E,p) \cap \{ j \leq r-j \}) \cap \{ u+v = 1 \}.$$

and that is equivalent to $J(\mathcal{D},E) = (\underline{y}+\lambda\underline{x}, \underline{z})$ with $\lambda = 0$.

(2.2.18) Corollary. If p is strongly normalized, $\delta = 1$ implies $e(E) = 2$.

(2.3) Preparation

(2.3.1) The main invariant used for proving the existence of a winning strategy is $\delta(\mathcal{D},E,p)$. In this paragraph the system of parameters p will be restricted a bit more in order to make controlable the behaviour of $\delta(\mathcal{D},E,p)$.

(2.3.2) In all this paragraph $(X,E,\mathcal{D},P)$ will denote a 4-upla of type 0-0 with $w(\mathcal{D},E) = 0$ and $p = (x,y,z)$ will be a strongly normalized regular system of parameters.

(2.3.3) Lemma. Let us suppose that $e(E) = 2$ and

(2.3.3.1) $$z_1 = z + \sum_{i \geq 2} \lambda_i x^i \; ; \quad y_1 = y$$

or that $e(E) = 1$ and

(2.3.3.2) $$z_1 = z + \sum_{i \geq 2} \lambda_i x^i \; ; \quad y_1 = y + \sum_{i \geq 2} \mu_i x^i .$$

Then $p_1 = (x,y_1,z_1)$ is a strongly normalized system of parameters.

Proof. p_1 is normalized since one does not change the expressions of the initial forms by making $z \longmapsto z_1$ $(y \longmapsto y_1)$ and one adds to the old coefficients terms of order high enough. On the other hand, one can see that the points in Exp(D,E,p) which contributes to (0,1) are not changed, actually the monomials remain the same ones, so p_1 is strongly normalized.

(2.3.4) <u>Definition</u>. $p = (x,y,z)$ is " prepared " iff one of the two following possibilities is satisfied

a) $\delta = \delta(\mathcal{D},E,p) \notin \mathbb{Z}$ or $\delta = 1$.

b) $\delta \in \mathbb{Z} - \{1\}$ and there is no change of the type

(2.3.4.1) $$y_1 = y + \mu x^{\delta} \; ; \; z_1 = z + \lambda x^{\delta}$$

with $\mu = 0$ if $e(E) = 2$, such that if $p_1 = (x,y_1,z_1)$ one has that

(2.3.4.2) $$\delta(\mathcal{D},E,p_1) > \delta \; .$$

(2.3.5) If p is not prepared, a change as in (2.3.4.1) will be called a "preparation" change for p. Thus we obtain p_1, if p_1 is not prepared, we may repeat. We have two possibilities: the algorithm stops in a step p_t which is prepared or the algorithm does not stop. In the last case, by composing all the changes we obtain

$$y^{\sim} = y + \sum \mu_i x^i \; ; \quad z^{\sim} = z + \sum \lambda_i x^i .$$

Before proving a result about $(x,y^{\sim},z^{\sim})$, we shall need the following lemma.

(2.3.6) <u>Lemma</u>. Let $n \geq \delta(\mathcal{D},E,p)$ and assume that $n \in \mathbb{Z}$ and $n \geq 2$. Let us consider the change

(2.3.6.1) $$y_1 = y + \sum_{i \geq n} \mu_i x^i \; ; \; z_1 = z + \sum_{i \geq n} \lambda_i x^i$$

where $\mu_i = 0$ for all i if $e(E) = 2$. Then

(2.3.6.2) $$\delta(\mathcal{D},E,p_1) \geq \delta(\mathcal{D},E,p)$$

where $p_1 = (x,y_1,z_1)$. Moreover one has the equality if $n \neq \delta(\mathcal{D},E,p)$.

<u>Proof</u>. Let us suppose that $\mathcal{D}$ is generated by

(2.3.6.3) $$D = ax\,\partial/\partial x + b\,\partial_y + c\,\partial/\partial z$$

with $\partial_y = \partial/\partial y$ or $y\partial/\partial y$ accordingly to e(E) = 1 or e(E) = 2. Then

(2.3.6.4) $$D = a_1 x_1 \partial/\partial x_1 + b_1 \partial_{y_1} + c_1 \partial/\partial z_1$$

where

(2.3.6.5) $$a_1 = a;\ b_1 = b + \sum_{i \geq n} i\mu_i x^i . a$$

$$c_1 = c + \sum_{i \geq n} i\lambda_i x^i a.$$

Let us observe that $m(\mathcal{D},E,p_1) \geq \min(n, m(\mathcal{D},E,p))$. From these equations and from the behaviour of the polygon of an hypersurface (|10|) one deduces that

(2.3.6.6) $$\Delta(\mathcal{D},E,p_1) \subset \Delta(\mathcal{D},E,p) + \mathbb{H}(n)$$

(see(2.2.8)) and thus the result.

(2.3.7) Proposition. In the situation of (2.3.5), one has that $\delta(\mathcal{D},E,p^\sim) = \infty$, where $p^\sim = (x, y^\sim, z^\sim)$ and thus $p^\sim$ is prepared.

Proof. If $\delta(\mathcal{D},E,p^\sim) < \infty$, there is t such that $\delta(\mathcal{D},E,p_t) < \delta(\mathcal{D},E,p^\sim)$. The passage $p_t \mapsto p^\sim$ is given by a change as (2.3.6.1) with strict inequality, and thus we have a contradiction.

(2.3.8) Definition. The algorithm defined in (2.3.5) will be called a "preparation of p".

(2.3.9) Corollary. There is always a prepared strongly normalized system of regular parameters (For short, we shall write p.s.n.s.r.p.).

(2.4) Main result

(2.4.1) The behaviour of $\delta(\mathcal{D},E,p)$ where p is a p.s.n.s.r.p. gives us the existence of a winning strategy.

(2.4.2) Proposition. If $(X,E\mathcal{D},P)$ is of the type 0-0, $w(\mathcal{D},E) = 0$ and $(X',E'\mathcal{D}',P')$

is a directional quadratic blowing-up of $(X,E,\mathcal{D},P)$, then one of the two following properties is satisfied

a) $\nu(\mathcal{D}',E',P') < r$ or $\dim \mathrm{Dir}(\mathcal{D}',E') = 0$

b) $e(E') \leq e(E)$.

Proof. It follows easily from the proof of (1.2.1).

(2.4.3) Remark. If $2 = e(E) \geq 1 = e(E')$ one has necessarily that $\delta(\mathcal{D},E,p) = 1$ for each strongly normalized r.s. of p. . Conversely, if $\delta(\mathcal{D},E,p) = 1$ for a strongly normalized r.s.of p. (hence $e(E)=2$) then one has that $e(E')=1$. (This also follows from the proof of (1.2.1))..

(2.4.5) Proposition. Let us suppose that $(X,E,\mathcal{D},P)$ is of the type 0-0, let $w(\mathcal{D},E) = 0$, let $p = (x,y,z)$ be a p.s.n.s.r.p. and let $(X',E',\mathcal{D}',P')$ be a directional quadratic blowing-up of $(X,E,\mathcal{D},P)$. Then one of the following properties is satisfied

a) $\nu(\mathcal{D}',E',P') < r$ or $\dim \mathrm{Dir}(\mathcal{D}',E') = 0$.

b) $e(E') < e(E)$.

c) $e(E') = e(E)$ and there exist a p.s.n.s.r.p. $p' = (x',y',z')$ for $(X',E',\mathcal{D}',P')$ such taht

(2.4.5.1) $$\delta(\mathcal{D}',E',p') = \delta(\mathcal{D},E,p) - 1.$$

Proof. We shall suppose that a) and b) are not satisfied. Then we have that the blowing-up is given by (T-1,0): $x = x'$; $y = x'y'$; $z = x'z'$. It is possible to obtain that $p = (x',y',z')$ works for c). First, if p' is a p.s.n.s.r.p. then one has that

(2.4.5.2) $$\Delta(\mathcal{D}',E',p') = \sigma(\Delta(\mathcal{D},E,p))$$

where $\sigma(u,v) = (u+v-1,v)$. This result follows from the definition of the polygon, the behaviour of the polygon of an hypersurface (|10|) and the fact that

(2.4.5.3) $$m(\mathcal{D}',E',p') = m(\mathcal{D},E,p) - 1.$$

Now (2.4.5.1) follows from (2.4.5.2) and the fact that $(0,1) \in \Delta(\mathcal{D},E,p)$. Thus it is enough to prove that p' is a p.s.n.s.r.p. Moreover, one has that

(2.4.5.4) $$\delta(\mathcal{D},E,p) \geq 2$$

since reasoning like for (2.4.5.2) and (2.4.5.3) one would have $\nu(\mathcal{D}',E',P') < r$ if (2.4.5.4) is not satisfied. We shall distinguish two cases:

Case A: $e(E) = 1$. Now we distinguish two possibilities. If $\delta(\mathcal{D},E,p) > 2$, then reasonning like for (2.4.5.2) and (2.4.5.3) one deduces easily that p' is strongly normalized. Moreover, if $\delta' = \delta(\mathcal{D},E,p)-1 \in \mathbb{Z}$ and the change $y'_1=y'+\mu x'^{\delta'}$, $z'_1=z'+\lambda x'^{\delta'}$, increases δ', then

(2.4.5.5) $$y_1 = y + \mu x^{\delta'+1} \; ; \; z_1 = z + \lambda x^{\delta'+1}$$

increases $\delta(\mathcal{D},E,p)$, which is a contradiction. Thus p' is a p.s.n.s.r.p. If $\delta' \notin \mathbb{Z}$ there is no problem. Let us suppose now that $\delta' = 1$. Since $(0,1) \in \Delta(\mathcal{D}',E',p')$, if p' is not strongly normalized then $J(\mathcal{D}',E')=(y'+\mu x',z'+\lambda x')$, $(\mu,\lambda) \neq (0,0)$. If

(2.4.5.6) $$y_1 = y + \mu x^2 \; ; \; z_1 = z + \lambda x^2$$

then $p_1 = (x,y_1,z_1)$ continues to be a p.s.n.s.r.p. such that $2 = \delta(\mathcal{D},E,p_1) = \delta(\mathcal{D},E,p)$ and it is enough to interchange p by p_1.

Case B: $e(E) = 2$. If $\delta(\mathcal{D},E,p) > 2$ we finish as above. If $\delta' = 1$ and p' is not strongly normalized, since $J(\mathcal{D},E) = (\underline{y},\underline{z})$ then we have (2.4.5.6) too. Now, if we make $z_1 = z + \lambda x^2$, we finish as above.

(2.4.6) <u>Definition</u>. Let $(X,E,\mathcal{D},P)$ be of the type 0-0. If $w(\mathcal{D},E) = 1$, we shall put

(2.4.6.1) $$\delta(\mathcal{D},E) = 0.$$

If $w(\mathcal{D},E) = 0$, we shall put

$$\delta(\mathcal{D},E) = \min\{\delta(\mathcal{D},E,p);\ p \text{ is a p.s.n.r.s.p.}\}.$$

(2.4.7.) <u>Theorem</u>. There is a strongly winning strategy for the reduction game of type 0-0.

Proof. The strategy is defined at the beginning of the section. If at the status t there is no permissible curve, then one has $\delta(\mathcal{D},E) < \infty$. If player B do not miss, then he must to choose the closed point given by the directrix, then results above (2.1.5), (2.4.5) shows that

$$(2.4.7.1) \qquad (w', e', \delta') \leq (w, e, \delta - 1)$$

for the lexicographic order, where w,e, δ are the invariants for $(X, E\mathcal{D}, P)$ and w', e', δ' are for $(X', E', \mathcal{D}', P')$, the corresponding quadratic transform. Now, in view of (1.2.3), we have a winning strategy. On the other hand actually it is a strongly winning strategy because there exists clearly a longest realization of the game.

3. INVARIANTS ASSOCIATED TO THE TYPE 0-1

(3.1) Polygons for type 0-1

(3.1.1) In the sequel we shall suppose that $(X,E,\mathcal{D},P)$ is of the type 0-1.

(3.1.2) Definition. A system of regular parameters p = (x,y,z) is "normalized" iff p is suited for (E,P), $\underline{y} \notin J(\mathcal{D},E)$ and E is given by x or by xy.

(3.1.3) Remark. A s. of r.p. as in (1.1.4) is always normalized, but the converse is not true, thus "normalized" has a weaker sense than the corresponding concept in the type 0-0.

(3.1.4) Let p = (x,y,z) be normalized and let us suppose that $\mathcal{D}$ is generated by

$$(3.1.4.1) \qquad D = ax\partial/\partial x + b\partial y + c\partial/\partial z$$

($\partial y = \partial/\partial y$ if E is given by x and $\partial y = y\partial/\partial y$ if E is given by xy). The "cloud of points" Exp (D,E,p) is defined as in (2.2.4). And the invariant $m(\mathcal{D},E,p)$ is defined as in (2.2.6). Finally, the polygon $\Delta(\mathcal{D},E,p)$ is defined as in (2.2.8).

(3.1.5) Remark. The points of

(3.1.5.1) $$\psi(\mathrm{Exp}\ (\mathcal{D},E,p) \cap \{(h,i,j);\ j \leq r-1\})$$

(see 2.2.8.1) are contained in $(\mathbb{Z}/r!)^2$. It follows that $\Delta(\mathcal{D},E,p)$ has only a finite number of vertices and they satisfy the above property (see |10|).

(3.1.6) Definition. Let us denote by

(3.1.6.1) $$(\alpha(\mathcal{D},E,p),\ \beta(\mathcal{D},E,p))$$

the coordinates of the vertex of lowest abscissa of $\Delta(\mathcal{D},E,p)$. This vertex will be called "main vertex". Let us denote by

(3.1.6.2) $$\varepsilon(\mathcal{D},E,p)$$

the value -1/slope of the segment joining the first and second vertices of $\Delta(\mathcal{D},E,p)$. If there is only one vertex, we put $\varepsilon(\mathcal{D},E,p) = \infty$.

(3.1.7) Remark. As for the case of surfaces, the numbers $(\beta,\varepsilon,\alpha)$ will be the main invariants for the control of the type 0-1 (see |10|). But there are certain differences, first the preparation algorithms are a bit more complicated and second the number of components e(E) will be added as an invariant.

(3.1.8) Proposition. Let p = (x,y,z) be a normalized system of regular parameters. Let Y be given by (x,z) and let Z be given by (y,z). Then Y is a permissible center, resp. Z is a permissible center, iff the polygon $\Delta(\mathcal{D},E,p)$ does not intersect the line u=1, resp. v=1 (u,v being the coordinates in $\mathbb{R}^2$).

Proof. If e(E) = 2 it follows easily from the standard results on hypersurfaces (see |10|). If e(E) = 1, the only problem is in the fact that

(3.1.8.1) $$\psi(\mathrm{Exp}\ (zb/y) \cap (j \leq r-1) \subset (v \geq 1)$$

may not imply that b has multiplicity r along Z. But in this case, necessarily $m(\mathcal{D},E,p) < \infty$ and v=1 must intersect the polygon.

(3.1.9) Proposition. Let $p = (x,y,z)$ be a normalized system of parameters and let Y, resp. Z, be given by (x,z), resp (y,z). Let $(X',E',\mathcal{D}',P')$ be a directional blowing-up of $(X,E\mathcal{D},P)$ such that it is of the type 0-1 and one of the following statements is satisfied:

a) Y is permissible, Y is the center of the blowing-up and the equations of the blowing-up are given by T-3.

b) Z is permissible, Z is the center of the blowing-up and the equations of the blowing-up are given by T-4.

c) Y and Z are not permissible , the blowing-up is quadratic and the equations are given by (T-1,0) or T-2.

Now, let $p' = (x',y',z')$ be given by the equations of the blowing-up. Moreover, let us suppose that p' is normalized. Then one has that

(3.1.9.1) $$(\beta(\mathcal{D}',E',p'),e(E'),\varepsilon(\mathcal{D}',E',p'),\alpha(\mathcal{D}',E',p')) <$$

(3.1.9.2) $$< (\beta(\mathcal{D},E,p),e(E),\varepsilon(\mathcal{D},E,p,),\alpha(\mathcal{D},E,p)).$$

for the lexicographic order.

Proof. If $e(E) = 2$, then $e(E') = 2$ and

(3.1.9.3) $$\Delta(\mathcal{D}',E',p') = \sigma(\Delta(\mathcal{D},E,p))$$

where

(3.1.9.4) $$\begin{aligned}\sigma(u,v) &= (u+v-1,v) \text{ if } (T-1,0)\\ &= (u,u+v-1) \text{ if } T-2\\ &= (u-1,v) \quad \text{if } T-3\\ &= (u,v-1) \quad \text{if } T-4\end{aligned}$$

then the result follows as in (|10|). Let us suppose that $e(E) = 1$. Let us denote by D,D' generators of $\mathcal{D}$ and $\mathcal{D}'$ obtained one from another by the equations of (I. 2.2.5). Now, with notations as in 2.2.7, one can see that the convex hull of

(3.1.9.5) $$\psi(\mathrm{Exp}(D',E',p') \cap (j \leq r-1)) + \mathbb{R}_0^2$$

is the image by σ of the convex hull of

(3.1.9.6) $$\psi(\mathrm{Exp}(D,E,p) \cap (j \leq r-1)) + \mathbb{R}_o^2.$$

Now, the result follows from the fact that if T-2 or T-4 then $m(\mathcal{D}',E',p') = \infty$, but β decreases. If (T-1,0)

(3.1.9.7) $$m(\mathcal{D}',E',p') = m(\mathcal{D},E,p) - 1$$

and if T-3

(3.1.9.8) $$m(\mathcal{D}',E',p') = m(\mathcal{D},E,p),$$

thus in cases (T-1,0) and T-3 one has (3.1.9.3).

(3.2.) Good preparation. First cases

(3.2.1) Here the parameters considered will be restricted in order to obtain a certain stability under transformations like the ones in (3.1.9).

(3.2.2) Although for the purposes of the existence of a winning strategy type 0-1 is treated as an unified type, there are significant differences in the technical behaviour of the cases $e(E) = 1$ and $e(E) = 2$.

(3.2.3) Lemma. Let $p = (x,y,z)$ be a normalized base and let us suppose that $e(E)=2$. Let us suppose that (s,t) is a vertex of $\Delta(\mathcal{D},E,p)$ and $(s,t) \in \mathbb{Z}_o^2$. Let us consider the change of coordinates

(3.2.3.1) $$z_1 = z + \lambda x^s y^t$$

where $\lambda \in k$. Let $p_1 = (x,y,z_1)$. Then one has that

a) p_1 is normalized.

b) $\Delta(\mathcal{D},E,p_1) \subset \Delta(\mathcal{D},E,p)$ and every vertex of $\Delta(\mathcal{D},E,p)$ different from (s,t) is a vertex of $\Delta(\mathcal{D},E,p_1)$.

Proof. a) trivial.

b) Let us suppose that $\mathcal{D}$ is generated by

(3.2.3.2) $$D = ax\partial/\partial x + by\partial/\partial y + c\partial/\partial z = $$
$$= a_1 x\partial/\partial x + b_1 y\partial/\partial y + c_1\partial/\partial z_1$$

Then one has that

(3.2.3.3) $$a = a_1;\ b = b_1;\ c_1 = c + \lambda s x^s y^t a + \lambda t x^s y^t b.$$

Now, let us fix a monomial which appears in a coefficient of D.

(3.2.3.4) $$M = \mu x^h y^i z^j.$$

We shall distinguish two cases. First, let us suppose that M does not define a point of

(3.2.3.5) $$\mathrm{Exp}\ (\mathcal{D},E,p) \cap (j \leq r-1).$$

Then M produces monomials, after (3.2.3.1), which eventually contributes to a point of $\Delta(\mathcal{D},E,p')$ in

(3.2.3.6) $$(s,t) + \mathbb{R}_o^2.$$

Moreover, if they contribute to (s,t), then $M = \mu z^r$ and it is a monomial of the coefficient c. Now, let us suppose that M contributes to a point $(s',t') \in \Delta\ (\mathcal{D},E,p)$. Then, if $(s',t') = (s,t)$, M produces after (3.2.3.1) monomials which contributes all of them to (s,t). If $(s',t') \neq (s,t)$, then it is produced a monomial

(3.2.3.7) $$\mu x^h y^i z_1^j$$

in the same coefficient as M and the rest of the monomials produced contribute to points which are in the segment joining (s,t) and (s',t') and differents from (s,t) and (s',t'). Now, b) follows easily.

(3.2.4) <u>Remark</u>. The expression "M contributes to a point (s,t) of $\Delta(\mathcal{D},E,p)$" may be explained by saying that M defines a point in $\mathrm{Exp}\ (\mathcal{D},E,p) \cap (j \leq r-1)$ and that the image by ψ of this point is (s,t). Let us observe that since contributions of different monomials never kill, it is possible that after (3.2.3.1) two different monomials would produce a pair of monomials which kill.

(3.2.5) Remark. From the proof of the lemma, one see that (s,t) is the only vertex of $\Delta(\mathcal{D},E,p)$ which may be is not in $\Delta(\mathcal{D},E,p_1)$ and this is a direct consequence of the existence of a monomial μz^r in the third coefficient.

(3.2.6) Definition. Let $p = (x,y,z)$ be normalized and let us suppose $e(E) = 2$. We shall say that a vertex (s,t) of $\Delta(\mathcal{D},E,p)$ is "well prepared" if one has one of the following possibilities:

a) $(s,t) \notin \mathbb{Z}_0^2$.

b) $(s,t) \in \mathbb{Z}_0^2$ and there in no change as (3.2.3.1) in such a way that

(3.2.6.1) $$\Delta(\mathcal{D},E,p_1) \subset \Delta(\mathcal{D},E,p) - \{(s,t)\}.$$

(3.2.7) Proposition. Let $p = (x,y,z)$ be normalized, $e(E) = 2$ and let us consider the change of coordinates

(3.2.7.1) $$z_1 = z + \sum_{(s,t)} \lambda_{st} x^s y^t$$

such that if $\lambda_{st} \neq 0$ then $(s,t) \in \Delta(\mathcal{D},E,p)$ and it is not a vertex, then one has that

a) $p_1 = (x,y,z_1)$ is normalized.

b) $\Delta(\mathcal{D},E,p) = \Delta(\mathcal{D},E,p_1)$.

Proof. It is similar to the proof of (3.2.5)

(3.2.8) Corollary. If $e(E) = 2$, then there exists always a normalized regular system of parameters $p = (x,y,z)$ such that every vertex of $\Delta(\mathcal{D},E,p)$ is prepared. Moreover, such a system may be obtained by a sequence of changes of the type (3.2.3.1) from an arbitrary normalized system.

Proof. We begin with a normalized system and make successively changes as (3.2.3.1) in each not prepared vertex. The limit of this convergent sequence satisfies that all the vertices are prepared, as a consequence of (3.2.7).

(3.2.9) Definition. Let us suppose that $e(E) = 2$ and $p = (x,y,z)$ is a normalized base. A "good preparation for p" is a sequence of coordinate changes as described in (3.2.8). If p_1 is the obtained base, we shall write $p \xrightarrow{w.p.} p_1$.

(3.3) <u>Good preparation e(E) = 1</u>

(3.3.1) Assume that $(X,E,\mathcal{D},P)$ is of the type 0-1 and that $e(E) = 1$.

(3.3.2) Before defining good preparation, let us make a study of the effect of a coordinate change of the type

(3.3.2.1) $$z_1 = z + \xi x^\alpha y^\beta$$

over the polygon. Let $p = (x,y,z)$ be a normalized base. Let

(3.3.2.2) $$D = ax\partial/\partial x + b\partial/\partial y + c\partial/\partial z$$

be a generator of $\mathcal{D}$. For a coefficient, vgr "a", we shall denote by Mon (a,p) the set of the monomials which appear in the expression of "a" as a series in x,y,z over k. We shall denote

(3.3.2.3) $$\text{Mon }(D,p) = \text{Mon }(a,p) \uplus \text{Mon }(b,p) \uplus \text{Mon }(c,p). \quad \text{(disjoint union)}$$

Let us suppose now that $M \in \text{Mon }(D,p)$. Let

(3.3.2.4) $$D_M = a'x\partial/\partial x + b'\partial/\partial y + c'\partial/\partial z$$

where a' (resp. b', resp. c')=0 if $M \notin \text{Mon}(a,p)$ (resp. Mon(b,p), resp. Mon(c,p)). and = M otherwise. One has that

(3.3.2.5) $$D = \sum_{M \in \text{Mon}(D,p)} D_M.$$

For each D_M, we shall define Exp, Δ, m, etc..., formally exactly in the same way as in (3.1).

(3.3.3) Let $p_1 = (x,y,z)$ with z_1 as in (3.3.2.1) and assume $(\alpha,\beta) \neq (0,0)$, p_1 is a normalized base too. Let us take

(3.3.3.1) $$M = \lambda x^h y^i z^j \in \mathrm{Mon}\ (D,p)$$

We are interested in the part of the set

(3.3.3.2) $$\mathrm{Mon}\ (D_M,p_1)$$

which gives in Exp (D_M,E,p_1) points in the set

(3.3.3.2) $$\{(\gamma,\delta,\varepsilon);\quad \varepsilon \leq r-1\} \cup \{(n,-1,r);\ n \in \mathbb{Z}\}.$$

and in the possible projections to $\Delta(D_M,E,p_1)$ and contributions to $m(D_M,E,p_1)$.

Let us suppose now that $M \in \mathrm{Mon}(a,p)$. One has that

(3.3.3.3) $$D_M = \lambda \sum_{s=0}^{j} \binom{j}{s} \xi^s [x^{h+s\alpha} y^{i+s\beta} z_1^{j-s} . \partial/\partial x + x^{h+(s+1)\alpha} y^{i+(s+1)\beta} z_1^{j-s} \partial/\partial z_1].$$

Assume first that $j \geq r-1$ (i.e. M does not produce any point of the polygon $\Delta(\mathcal{D},E,p)$) Then the points induced in (3.3.3.2) do not contribute to "m" and they project over points in

(3.3.3.4) $$[(\alpha,\beta) + \mathbb{R}_0^2] - \{(\alpha,\beta)\}.$$

Assume now $j \leq r-2$. Then M contributes to a point $(\alpha',\beta') = (h/(r-j-1),\ i/(r-j-1))$ in $\Delta(\mathcal{D},E,p)$ all the monomials in Mon (D_M,p_1) induce points in (3.3.3.2) and none contributes to "m". We shall distinguish two cases: $(\alpha,\beta) = (\alpha',\beta')$ or not. If $(\alpha,\beta) = (\alpha',\beta')$ then all the points contribute under projection to $(\alpha,\beta) \in \Delta(D_M,E,p_1)$. If $(\alpha,\beta) \neq (\alpha',\beta')$ then all the points contribute to points in $\Delta(D_M,E,p_1)$ placed in the segment joining (α,β) and (α',β'), none of them contributes to (α,β) and the only one which contributes to (α',β') corresponds to the monomial

(3.3.3.5) $$\lambda x^h y^i z_1^j \in \mathrm{Mon}\ (D_M,p_1)$$

as monomial of the first coefficient. (The study of this case is similar for $e(E) = 2$).

Let us suppose that $M \in \mathrm{Mon}\ (b,p)$. Then one has

(3.3.3.6) $$D_M = \lambda \sum_{s=0}^{j} \xi^s \binom{j}{s} [x^{h+s\alpha} y^{i+s\beta} z_1^{j-s} \partial/\partial y + \beta x^{h+(s+1)\alpha} y^{i-1+(s+1)\beta} z_1^{j-s} \partial/\partial z].$$

Let us suppose first that $j \geq r$ (i.e. M does not produce any point of $\Delta(\mathcal{D},E,p)$ and it does not contribute to $m(\mathcal{D},E,p)$). Assume first $\beta \geq 1$. Then, there is no contribution to "m" and the points in (3.3.3.2) project over points of $\Delta(D_M,E,p_1)$ placed in (3.3.3.4), except for $(\alpha,\beta) = (0,1)$, $M = \lambda z^r$, in this last case all this points contribute to (α,β). Let us suppose that $\beta = 0$, if $i \neq 0$, there is no contribution to "m" and the points in (3.3.3.2) project over points of $\Delta(D_M,E,p_1)$ placed in (3.3.3.4). If $i = 0$, then the contribution to "m" is given only by the monomial

(3.3.3.7) $$\lambda \xi^s \binom{j}{s} x^{h+s} z_1^{r-1}, \quad s = j-r+1$$

(if the coefficient is nonnull) and we have

(3.3.3.8) $$m(D_M,E,p_1) = h + \alpha(j-r+1) \geq \alpha .$$

Moreover, the points in (3.3.3.2) project over points of $\Delta(D_M,E,p_1)$ placed in

(3.3.3.9) $$[(\alpha,\beta) + \mathbb{H}(h+\alpha(j-r+1))] - \{(\alpha,\beta)\}$$

(see (2.2.7) for notation). Assume now that $j = r-1$. If $i \neq 0$, there is no contribution to "m" and the points in (3.3.3.2) project over points of $\Delta(D_M,E,p_1)$ placed in (3.3.3.4) except for $\beta = 0$, $M = \lambda y z^{r-1}$, in this case all this points contribute to (α,β). If $i = 0$, then the contribution to "m" is given by monomial $\lambda x^h z_1^{r-1}$ and

(3.3.3.10) $$m(D_M,E,p_1) = h \geq m(\mathcal{D},E,p).$$

The points in (3.3.3.2) project over points of $\Delta(D_M,E,p_1)$ placed in

(3.3.3.11) $$[(\alpha,\beta) + \mathbb{H}(h)] - \{(\alpha,\beta)\}.$$

Finally, assume $j < r-2$. Let $(\alpha',\beta') = (h/(r-j-1), (i-1)/(r-j-1))$ be the point of $\Delta(\mathcal{D},E,p)$ induced by M. In this case we have the same result as in the case $M \in \mathrm{Mon}(a,p)$.

Let us suppose that $M \in$ Mon (c,p). Then, there is never contribution to "m". If $j \geq r$, the points in (3.3.3.2) project over points of $\Delta(D_M,E,p_1)$ placed in (3.3.3.4), except for $M = \lambda z^r$, which contributes only to (α,β). If $j \leq r-1$ and (α',β') is the corresponding point in $\Delta(\mathcal{D},E,p)$, we have results as above.

(3.3.4) Remark. To obtain $\Delta(\mathcal{D},E,p_1)$ one has to make the sum of (3.3.2.5) and to consider only those monomials which do not kill in this sum.

(3.3.5) Lemma. Let $p = (x,y,z)$ be a normalized base, let (α_i,β_i), $i = 1,\ldots,t$ be the first t vertices of the polygon $\Delta(\mathcal{D},E,p)$, let $(\alpha_t,\beta_t) \in \mathbb{Z}_o^2$ and assume that if $\beta_t = 0$ one has the additional property

(3.3.5.1) $$\alpha_i + \beta_i \cdot \alpha_t > \alpha_t \qquad i = 1,\ldots,t-1.$$

Let us denote by l_i the length of the segment joining (α_i,β_i) with the next vertex and let us denote by $-1/\varepsilon_i$ the slope of l_i. Let us consider the coordinate change

(3.3.5.2) $$z_1 = z + \lambda x^\alpha y^\beta$$

where $\alpha = \alpha_t$, $\beta = \beta_t$. Then one has that

a) $p_1 = (x,y,z_1)$ is normalized.

b) (α_i,β_i) , $i=1,\ldots,t-1$ are the t-1 first vertices of $\Delta(\mathcal{D},E,p_1)$.

c) The monomials in p_1 which contribute to the vertices (α_i,β_i), $i = 1,\ldots,t-1$ and to the points in the segments joining this vertices are formally the same as for p. Moreover, one has that z^r is in the third coefficient with respect to p iff it is so with respect to p_1.

d) If $t > 1$, then one has that

(3.3.5.3) $$(\varepsilon_{t-1},-l_{t-1}) \leq (\varepsilon'_{t-1},-l'_{t-1})$$

for the lexicographic order, where ε',l' means the values in $\Delta(\mathcal{D},E,p_1)$.

e) If $t = 1$, then one has that

$$(\alpha_1,\beta_1) \leq (\alpha'_1,\beta'_1)$$

for the lexicographic order, where (α'_1,β'_1) is the main vertex of $\Delta(\mathcal{D},E,p_1)$.

Proof. If follows from (3.3.3).

(3.3.6) Lemma. Let p = (x,y,z) be a normalized base and let us consider the change of coordinates

(3.3.6.1) $$z_1 = z + \sum_{(\alpha,\beta)} \lambda_{\alpha\beta} x^{\alpha} y^{\beta}.$$

Let Δ be the convex hull of

$$\{(\alpha,\beta);\ \lambda_{\alpha\beta} \neq 0\} + \mathbb{R}_0^2.$$

Then

a) $p_1 = (x,y,z_1)$ is normalized.

b) If w is the first vertex of Δ (looking in the sense of increassing abscissas) which is not in the polygon $\Delta(\mathcal{D},E,p)$, then w is a vertex of $\Delta(\mathcal{D},E,p_1)$.

c) If all the vertices of w are contained in $\Delta(\mathcal{D},E,p)$ and not in the "border"

(3.3.6.2) $$\partial(\Delta(\mathcal{D},E,p)) := \cup\{\text{segments of finite length joining vertices of } \Delta(\mathcal{D},E,p)\}$$

and the vertex $w_0 = (n,0)$ of Δ (if it exists) satisfy that

(3.3.6.3) $$\alpha + \beta n > n$$

for each vertex (α,β) of $\Delta(\mathcal{D},E,p)$ such that $\beta \geq 0$. Then $\Delta(\mathcal{D},E,p_1)$ has exactly the same vertices as $\Delta(\mathcal{D},E,p)$ until those of ordinate strictly negative.

Proof. If is enough to make computations as in (3.3.3) for (3.3.6.1) and to consider the sum of (3.3.2.5) after the change. For b), note that there always exist monomials in D which produce the vertex w after the change, since we have type 0-1, and p is normalized.

(3.3.7) Definition. Let p = (x,y,z) be a normalized base and let (α_i,β_i), i=1,...,t be the first t vertices of the polygon $\Delta(\mathcal{D},E,p)$. We shall say that $\Delta(\mathcal{D},E,p)$ is "well prepared" until the vertex (α_t, β_t) iff for each (α_i, β_i) one has one of the following properties:

a) $(\alpha_i, \beta_i) \notin \mathbb{Z}_o^2$.

b) $(\alpha_i, \beta_i) \in \mathbb{Z}_o^2$, $\beta_i \neq 0$ and there is no change of the type $z_1 = z + \lambda x^{\alpha_i} y^{\beta_i}$ which migth increase

$$(\alpha_1, \beta_1, \varepsilon_1, -1_1, \ldots, \varepsilon_{i-1}, -1_{i-1}, \varepsilon_i, -1_i) \qquad (3.3.7.1)$$

for the lexicographic order.

c) $(\alpha_i, \beta_i) \in \mathbb{Z}_o^2$, $\beta_i = 0$ and there exist (α_j, β_j) with $j < i$ such that $\alpha_j + \beta_j \cdot \alpha_i \leq \alpha_i$.

d) $(\alpha_i, \beta_i) \in \mathbb{Z}_o^2$, $\beta_i = 0$, c) is not true and no change $z_1 = z + \lambda x^{\alpha_i}$ may increase (3.3.7.1). We shall say that $\Delta(\mathcal{D}, E, p)$ is "well prepared" iff it is well prepared until the last vertex.

(3.3.8) Theorem. There exists always a normalized base $p = (x,y,z)$ for which $\Delta(\mathcal{D}, E, p)$ is well prepared.

Proof. Let $p' = (x',y',z')$ be a normalized base. Let (α'_1, β'_1) be the first vertex of $\Delta(\mathcal{D}, E, p')$. If $\Delta(\mathcal{D}, E, p')$ is well prepared until (α'_1, β'_1), we do nothing. If not, we make the change $z'_1 = z' + \lambda x'^{\alpha'_1} y'^{\beta'_1}$ which increases (3.3.7.1) the most. We repeat. In this way we obtain a convergent change of coordinates

$$z'_1 = z' + \sum \lambda_{\alpha\beta} x'^{\alpha} y'^{\beta} \qquad (3.3.8.1)$$

By applying lemmas (3.3.5) and (3.3.6) we conclude that $\Delta(\mathcal{D}, E, p'_1)$ is well prepared until the first vertex, $p'_1 = (x',y',z'_1)$. Now we repeat the algorithm with the second vertex, and so on. The composite of all the changes made gives us the desired normalized base.

(3.3.9) Remark. If $p = (x,y,z)$ is a normalized base such that $(0,1) \notin \Delta(\mathcal{D}, E, p)$, then if

$$D = ax\partial/\partial x + b\partial/\partial y + c\partial/\partial z$$

generates $\mathcal{D}$, one has that $(0,1,r-1) \notin \mathrm{Exp}\,(b,p)$, since otherwise $\mathcal{D}$ could not be of the type 0-1. Moreover, if p is such that $\Delta(\mathcal{D},E,p)$ is well prepared until the first vertex, one has that $(0,1) \notin \Delta(\mathcal{D},E,p)$, since for each normalized base there is a change $z_1 = z + \lambda y$ such that $J^r(c_1) = (\underline{z}_1)$ (c_1 = coeff. of $\partial/\partial z_1$) and if (0,1) continues to be in $\Delta(\mathcal{D},E,p)$, type 0-1 is not possible.

(3.3.10) Definition. The chage $p' \longrightarrow p$ of the proof of (3.3.8) is a "good preparation change" of p' and we shall denote it by $p' \xrightarrow{w.p.} p$.

(3.4) Very good preparation

(3.4.1) As we shall see later (3.5), the systems of parameters for which the polygon is well prepared have nice behaviours under transformations of the type (T-1,0) T-2, T-3 or T-4. But one has to ameliorate the choice of the parameters in order to control the singularities under T-1, ζ with $\zeta \neq 0$.

(3.4.2) In this paragraph we shall suppose that e(E) = 1 and $(X,E,\mathcal{D},P)$ is of the type 0-1.

(3.4.3) Proposition. Let p = (x,y,z) be a normalized base and let us consider the coordinate change

(3.4.3.1) $$y_1 = y + \zeta x^n, \qquad \zeta \in k$$

Let $p_1 = (x,y_1,z)$. Then

a) p_1 is normalized.

b) $\Delta(\mathcal{D},E,p_1) + \mathbb{H}(n) = \Delta(\mathcal{D},E,p) + \mathbb{H}(n)$.

c) Let (α_i,β_i), i=1,...,t be the vertices of

$$\Delta(\mathcal{D},E,p) + \mathbb{H}(n).$$

Then, the monomials in p_1 which contribute to (α_i,β_i) i=1,...,t and to the points in the segments joining this vertices are formally the same as for p. Moreover, one has that μz^r is in the third coefficient with respect to p iff it is so with respect to p_1. (All this is with reference to a fixed generator of $\mathcal{D}$).

d) For each $j \leq t$, $\Delta(\mathcal{D},E,p_1)$ is well prepared until (α_j,β_j) iff it is so for $\Delta(\mathcal{D},E,p)$.

e) If $\Delta(\mathcal{D},E,p)$ is well prepared with respect to (α_1,β_1), then one has that

(3.4.3.2) $$m(\mathcal{D},E,p_1) \geq \min(n+1, m(\mathcal{D},E,p)).$$

<u>Proof</u>. a) is trivial and c) $\Rightarrow$ d). Let $M = \lambda x^h y^i z^j$ be a monomial in Mon (D,p) for a fixed generator of $\mathcal{D}$:

(3.4.3.3) $$D = ax\,\partial/\partial x + b\,\partial/\partial y + c\,\partial/\partial z.$$

If $M \in$ Mon (a,p), let $j \geq r$, then there is no contribution. Let $j = r-1$, then it possibly contributes to "m" with the monomial

(3.4.3.4) $$\lambda\, n\,\zeta^{i+1} x^{h+n(i+1)} z^{r-1}$$

and thus (if $n\zeta^{n+1} \neq 0$) one has

(3.4.3.4) $$m(D_M,E,p_1) = h+n(i+1)$$

If M corresponds to a point (α,β), then after (3.4.3.1) it contributes in $\Delta(D_M,E,p_1)$ to points of the type

(3.4.3.5) $$(\alpha,\beta) + \mu(n,-1) \qquad \mu \geq 0 \qquad \mu \in \mathbb{Q}.$$

and the monomial which gives (α,β) is formally M.

If $M \in$ Mon (b,p) and $j \geq r$ there is no contribution, if $j = r-1$, there is a contribution to "m" by $\lambda\zeta^i x^{h+n.i} z^{r-1}$ and then

(3.4.3.6) $$m(D_M,E,p_1) = h+n.i.$$

If $j \leq r-2$ we have a result as above.

If $M \in$ Mon (c,p) and $j \leq r$, nothing occurs, if $j \leq r-1$ we have a result as (3.4.3.5) above. This proves b) and c). Finally for e) it is enough to observe that in (3.4.3.6) one has always $i \geq 2$ (see 3.3.9).

(3.4.5) Definition. Let $p = (x,y,z)$ be a normalized base such that $\Delta(\mathcal{D},E,p)$ is well prepared. $\Delta(\mathcal{D},E,p)$ is "very well prepared" for p iff one has one of the following properties:

a) $\varepsilon = \varepsilon(\Delta(\mathcal{D},E,p)) \notin \mathbb{Z}_o$.

b) $\varepsilon \in \mathbb{Z}_o$ and for each change $y_1 = y + \xi x^{\varepsilon}$ followed by a good preparation $p_1 \mapsto p'$, where $p_1 = (x,y_1,z)$ one has that

$$(\varepsilon',-l') \leq (\varepsilon,-l) \qquad (3.4.5.1)$$

for the lexicographic order, where l denotes the length of the first segment of the polygon and "'" denotes "things in $\Delta(\mathcal{D},E,p')$".

(3.4.6) Proposition. There exists always a normalized very well prepared base.

Proof. Take a normalized well prepared base. If it is not very well prepared, make a change $y_1 = y + \zeta x^{\varepsilon}$ followed by a good preparation which decreases strictly $(\varepsilon,-l)$. Repeat. If we do not stop then the resulting polygon has only one vertex which is well prepared and thus the resulting base is very well prepared.

(3.4.7) Remark. The above algorithm does not assure that the pair $(\varepsilon,-l)$ is the same for the result of each realization of the algorithm, but this will not be used for the proof of the existence of a winning strategy.

4. A WINNING STRATEGY FOR TYPE 0-1

In this section we shall prove the results of stability of the systems of parameters well prepared and very well prepared which will allow us to profite of the result in (3.1.9) in order to use the invariant $(\beta,e,\varepsilon,\alpha)$ for the control of the singularities of the type 0-1.

(4.1) Good preparation stability

(4.1.1) We shall suppose always that $(X,E,\mathcal{D},P)$ is of the type 0-1. Let $p = (x,y,z)$ denote a normalized system of parameters such that $\Delta(\mathcal{D},E,p)$ is well prepared. Let us consider a directional blowing-up $(X',E',\mathcal{D}',P')$ which is quadratic only if (x,z) and (y,z) are not permissible and that is given by (T-1,0), T-2, T-3 or T-4 from p. and let us denote by $p' = (x',y',z')$ the resulting system of parameters. In this paragraph the phrase "player A has won" will mean that "$\nu(\mathcal{D}',E',P') < r=\nu(\mathcal{D},E,P)$, or dim Dir $(\mathcal{D}',E') = 0$, or $(X',E'\mathcal{D}',P')$ is of the type 0-0".

In all the paragraph we shall suppose that $\mathcal{D}$ is generated by

(4.1.1.1) $$D = ax\partial/\partial x + b\partial y + c\partial/\partial z$$

where ∂y means $\partial/\partial y$ if $e(E) = 1$ and $y\partial/\partial y$ if $e(E) = 2$, and that $\mathcal{D}'$ is generated by

(4.1.1.2) $$D' = a'x'\partial/\partial x' + b'\partial y' + c'\partial/\partial z'.$$

Our first result concerns to the stability of the equation of the directrix in a well prepared situation.

(4.1.2) Proposition. One has always $\underline{z} \in J^r(c)$.

Proof. Let us suppose that $\underline{z} \notin J^r(c)$. If $e(E) = 2$, since we have type 0-1, necessarily

(4.1.2.1) $$J^r(c) = (\underline{z} + \lambda\underline{x} + \mu\underline{y}).$$

If $\mu \neq 0$, (0,1) is the first vertex of $\Delta(\mathcal{D},E,p)$ and the coordinate change $z_1 = z + \mu y$ dissolves this vertex. If $\mu = 0$, $\lambda \neq 0$, then (1,0) is the last vertex and $z_1 = z + \lambda x$ dissolves it. Let us suppose that $e(E) = 1$. First, let us suppose that dim Dir $(\mathcal{D},E) = 2$, then

(4.1.2.2) $$J^r(\mathcal{D},E) = (\underline{z} + \lambda\underline{x} + \mu\underline{y}) = J_H(\mathcal{D},E)$$

where $H = H(\mathrm{Dir}(\mathcal{D},E))$, (see (1.1.2)). One has

(4.1.2.3) $$J_H(\mathcal{D},E) = J^r(D(z+\mu y))) = J^r(c+\mu b).$$

If $\mu \neq 0$, then there exists a monomial $\gamma.\underline{y}^r$, $\gamma \neq 0$, in the initial form of c or b. Thus (1,0) is the main vertex of $\Delta(\mathcal{D},E,p)$. If we make $z_1 = z + \mu y$, this vertex disappears and the sequence in (3.3.7.1) will be increased. If $\mu = 0$, $\lambda \neq 0$, then

(4.1.2.3) $$J_H(\mathcal{D},E) = J^r(c).$$

Moreover, the possibility c) of (3.3.7) is not possible, because there is no other vertices in $u+v = 1$, $v \geq 0$, now, the change $z_1 = z + \lambda x$ dissolves this vertex without touching the preceeding ones, contradiction. Let us suppose that dim Dir $(\mathcal{D},E) = 1$. Since one has type 0-1 and p is normalized, we have that

(4.1.2.4) $$J^r(\mathcal{D},E) = (\underline{x},\underline{z} + \lambda \underline{y})$$

and

(4.1.2.5) $$J_H(\mathcal{D},E) = J^r(D(z+\lambda y)) = J^r(c+\lambda b)$$

$$J^r(c+\lambda b) = (\underline{z}+\lambda \underline{y}+\mu \underline{x})$$

now, we can reason as above (note that this does not imply that there is no vertices in $u+v = 1$, but only that there is no vertices in $u+v = 1$, $v \geq 0$).

(4.1.3) Theorem. If $e(E) = 2$, then one of the two following possibilities is satisfied:

a) Player A has won.

b) $(X',E',\mathcal{D}',P')$ is of the type 0-1, p' is normalized and $\Delta(\mathcal{D}',E',p')$ is well prepared.

Proof. Let us suppose that a) is not true. The first two statements in b) have yet been proved. Let us suppose that $\Delta(\mathcal{D}',E',p')$ is not well prepared. Then, there exists a vertex $(\alpha',\beta') \in \Delta(\mathcal{D}',E',p')$ which may be dissolved by the coordinate change $z'_1 = z'+\lambda x'^{\alpha'} y'^{\beta'}$. Let $(\alpha,\beta) = \sigma^{-1}(\alpha',\beta')$ where σ is as in the proof of (3.1.9), (see (3.1.9.4)). Then (α,β) is a vertex of $\Delta(\mathcal{D},E,p)$ (see (3.1.9.3)) and if we make $z_1 = z+\lambda x^{\alpha} y^{\beta}$, then the hypothesis assures us that

(4.1.3.1) $$\Delta(\mathcal{D},E,p) = \Delta(\mathcal{D},E,p_1)$$

and that $\Delta(\mathcal{D},E,p_1)$ is w.p., $p_1 = (x,y,z_1)$. Moreover, one has that $\alpha+\beta > 1$, then the directional transform (T-1,0), T-2, T-3 or T-4 from p coincides with the one from p_1 (also the property for (x,z) or (y,z) of being permissible) and the base obtained from p_1 is exactly (x',y',z'_1). The contradiction appears by applying (3.1.9.3) to p_1 jointly with (4.1.3.1).

(4.1.4) Theorem. If e(E) = 1, then one of the following possibilities is satisfied:

a) Player A has won.

b) $(X',E',\mathcal{D}',P')$ is of the type 0-1, e(E') = 1, p' is normalized and $\Delta(\mathcal{D}',E',p')$ is well prepared.

c) $(X',E',\mathcal{D}',p')$ is of the type 0-1, e(E') = 2, p' is normalized and $\Delta(\mathcal{D}',E',p')$ is well prepared until the first vertex.

Proof. Let us suppose that a) is not true. First, let us suppose that e(E') = 1, i.e. we make (T-1,0) or T-3. Looking at the end of the proof of (3.1.9), we can reason just as in the proof of (4.1.3) above, but with the following remarks: take (α',β') the first not prepared vertex and observe that if c) of (3.3.7) is not true for (α',β') it is not true for (α,β). Let us suppose now that e(E') = = 2, i.e. we make T-2 or T-4. Let (α',β') be the first vertex, then, looking at the proof of (3.1.9) one has that $(\alpha',\beta') = \sigma((\alpha,\beta))$, where (α,β) is the first vertex of $\Delta(\mathcal{D},E,p)$ and σ is given by (3.1.9.4), now, we can reason as in (4.1.3) above.

4.2 Very good preparation stability

(4.2.1) The transformations $(T-1,\zeta)$, $\zeta \neq 0$, are considered. $(X,E,\mathcal{D},P)$ will be of the type 0-1 and p=(x,y,z) will be a normalized and well prepared base. Let $y_1=y+\zeta x$, the $(T-1,\zeta)$ from p is the same as (T-1,0) from $p_1=(x,y_1,z)$.

(4.2.2) After $(T-1,\zeta)$, e(E')=1. Thus by (3.1.9) one has only to control the ordinate β of the main vertex. A "virtual" transition to the case e(E)=1 is made before blowing-up, in the case e(E) = 2.

(4.2.3) Proposition. Let $e(E)=2$. Let E_o be such that $I(E_o) = (x)$ (then $E_o \subsetneq E$).Then $\mathcal{D} = (\alpha(\mathcal{D}),E_o)$, $(X,E_o,\mathcal{D},P)$ is of the type 0-1 and p is normalized and well prepared relatively to $(X,E_o,\mathcal{D},P)$.

Proof. The only possible common divisor of the new coefficients is "y" but this is not possible since $J(\mathcal{D},E) = (\underline{z})$, this proves $\mathcal{D} = (\alpha(\mathcal{D}),E_o)$. One has type 0-1 since the last coefficient remains unchanged and the middle coefficient has order r+1. Clearly p is normalized and it is well prepared since $\Delta(\mathcal{D},E,p) = \Delta(\mathcal{D},E_o,p)$.

(4.2.4) Corollary. With notation as above, if $p_1 = (x,y_1,z)$ is given as in (4.2.1), then

a) p_1 is normalized and $\Delta(\mathcal{D},E_o,p_1)$ is well prepared until the first vertex (which is the same as in $\Delta(\mathcal{D},E,p)$).

b) A well preparation of $\Delta(\mathcal{D},E_o,p_1)$ can be made by changes of the type $z_1 = z+\lambda x^\alpha y_1^\beta$ where $\alpha+\beta \geq 2$ (and thus the equations T-1,0 are not affected).

Proof. If follows from (3.4.3).

(4.2.5) Theorem. Let $e(E) = 1.$, let us suppose that (x,z) does not define a permissible center, and let $z \mapsto z_1$ be a good preparation of the polygon $\Delta(\mathcal{D},E,p_1)$ where $p_1 = (x,y_1,z)$ is as in (4.2.1). Let us denote $p_2 = (x,y_1,z_1)$. Then

a) The change $z \mapsto z_1$ is obtained from changes $z \mapsto z+\lambda x^\alpha y_1^\beta$, where $\alpha+\beta \geq 2$.

b) The first vertex of $\Delta(\mathcal{D},E,p_2)$ is the same as the first vertex of $\Delta(\mathcal{D},E,p)$.

c) $\varepsilon(\mathcal{D},E,p_2) \leq 1$ (see (3.1.6) for definition).

Proof. a) and b) follows from (3.4.3). If $\varepsilon(\mathcal{D},E,p) \leq 1$, c) is trivial by the definition of very good preparation. Thus, the only remaining case is

(4.2.5.1) $$\varepsilon(\mathcal{D},E,p) > 1.$$

We shall prove that in this case one has that

(4.2.5.2) $$\varepsilon(\mathcal{D},E,p_2) = 1.$$

Let us denote by (α,β) the first vertex of $\Delta(\mathcal{D},E,p)$. We shall distinguish three cases: $\alpha+\beta \notin \mathbb{Z}_0$, $(\alpha,\beta) \in \mathbb{Z}_0^2$ and $(\alpha,\beta) \notin \mathbb{Z}_0^2$ but $\alpha+\beta \in \mathbb{Z}_0$.

<u>First case</u>. $\alpha+\beta \notin \mathbb{Z}_0$. Thus $y \mapsto y_1$ generates a vertex of $\Delta(\mathcal{D},E,p_1)$ placed in the segment joining (α,β) with $(\alpha+\beta+1,-1)$ such that it is $(\alpha+\beta,0)$ or it has negative ordinate. Thus it is well prepared. As $\alpha < 1$ because of the assumption on (x,z), then the new vertex is different from (α,β) and so $\varepsilon(\mathcal{D},E,p_1) = 1$. Moreover, this situation cannot be modified by good preparation.

<u>Second case</u>. $(\alpha,\beta) \in \mathbb{Z}_0^2$. First, let us make some reductions. Let us suppose that $\mathcal{D}$ is generated by

(4.2.5.1) $$D = ax\partial/\partial x + b\partial/\partial y + c\partial/\partial z.$$

Since one has type 0-1, we can suppose that there is no monomials $\lambda yz^{r-1} \in \mathrm{Mon}\,(b,p)$. If $m(\mathcal{D},E,p) = 1$, clearly $\varepsilon(\mathcal{D},E,p) = 1$, thus there is not a monomial $\lambda xz^{r-1} \in \mathrm{Mon}\,(b,p)$ Now, looking at (3.3.3) and in view of the fact $(\alpha,\beta) \neq (0,1)$, we may suppose that

(4.2.5.2) $$\mathrm{Exp}\,(\mathbf{D},E,p) \subset L,$$

where L is the segment joining $(0,0,r)$ and $(r\alpha,r\beta,0)$, since the other monomials do not contribute for the result. Thus, we may suppose

$$a = \sum_{i=1}^{r-1} \lambda_{i,a} x^{\alpha(r-i)} y^{\beta(r-i)} z^{i-1}$$

(4.2.5.3) $$b = \sum_{i=1}^{r-1} \lambda_{i,b} x^{\alpha(r-i)} y^{\beta(r-i)+1} z^{i-1}$$

$$c = \lambda_{r,c} z^r + \sum_{i=0}^{r-1} \lambda_{i,c} x^{\alpha(r-i)} y^{\beta(r-i)} z^i, \qquad \lambda_{r,c} \neq 0$$

Let us suppose that $(\alpha+\beta,0)$ is a vertex of $\Delta(\mathcal{D},E,p_1)$ (as in the first case) but it is not well prepared, and let us suppose that a change $z_1 = z + \mu x^{\alpha+\beta}$ dissolves this vertex. Then

(4.2.5.4) $$\lambda_{i,a} = 0 \qquad i = 1,\ldots,r-1$$

since otherwise the vertex $(\alpha+\beta,0)$ could not have been eliminated. Now, if there is not a vertex of negative ordinate, one deduces that

(4.2.5.5) $$\lambda_{i,b} = 0 \qquad i = 1,\ldots,r-1$$

Now, we have only c. To simplify, let us suppose $\lambda_{r,c} = 1$. In order to eliminate $(\alpha+\beta,0)$, we must to consider only the part of c (after $y \mapsto y_1$) given by

(4.2.5.6) $$z^r + \sum_{i=0,\ldots,r-1} \lambda_{i,c}(-\zeta)^{\beta(r-1)} x^{(\alpha+\beta)(r-i)} z^i$$

if $(\alpha+\beta,0)$ disappears after $z \mapsto z_1$, necessarily one has

(4.2.5.7) $$\lambda_{i,c}(-\zeta)^{\beta(r-1)} = \binom{r}{i}\mu^{r-i} \qquad i = 0,\ldots,r-1$$

and then, we could have eliminated (α,β) "a priori" with a change $z=z_1+(\mu/(-\zeta)^{\beta})x^{\alpha}y^{\beta}$. Contradiction.

<u>Third case</u>. $(\alpha,\beta) \notin \mathbb{Z}_o^2$, $\alpha+\beta \in \mathbb{Z}_o$. We can reason as above, until (4.2.5.5) Moreover, to obtain the contradiction, we can suppose (4.2.5.7) too. Before starting, let us observe that necessarily $\alpha+\beta \geq 2$ and then $\beta \geq 1$ by the hypothesis. In fact, if $\alpha+\beta = 1$, then by the proof of 4.1.2, $\alpha > 1$ and this contradicts the hypothesis. Once made $y \longmapsto y_1$ and $z \longmapsto z_1 = z + \mu x^{\alpha+\beta}$, one can assume that the only (third) coefficient is

(4.2.5.8) $$z_1^r + \sum_{i=0}^{r-1} \sum_{j=0}^{(r-i)-1} \lambda_{i,c}(-\zeta)^j \binom{\beta(r-i)}{j} x^{\alpha(r-i)+j} y^{\beta(r-i)-j} (z_1 - x^{\alpha+\beta})^i.$$

From (4.2.5.7) we deduce that if $r \neq 0 \pmod{\tau}$, where τ is the characteristic of the field k, then $\lambda_{r-1,c} \neq 0$ and thus

(4.2.5.9) $$\beta(r-(r-1)) = \beta \in \mathbb{Z}_o$$

contradiction with the hypothesis of this case. Thus, necessarily $r=0 \pmod{\tau}$ (this implies $\tau > 0$). Let $l \in \mathbb{Z}_o$ such that τ^l divide r and τ^{l+1} does not divide r. Then one has that

(4.2.5.10) $$r - \tau^{l} = \max (i, \lambda_{i,c} \neq 0),$$

in view of (4.2.5.7). Let $i_o = r-\tau^l$. Since $\beta \notin \mathbb{Z}_o$, then τ^l does not divide $\beta(r-i_o) = \beta\tau^l$. Let l' be such that $\tau^{l'}$ divides $\beta\tau^l$ and $\tau^{l'+1}$ does not divide $\beta\tau^l$. One has that $l' < l$ and if $j_o = \beta\tau^l - \tau^{l'}$, then

(4.2.5.11) $$\binom{\beta(r-i_o)}{j_o} \neq 0 \pmod{\tau}.$$

Then, the coefficient in (4.2.5.8) of

(4.2.5.12) $$x^{\alpha(r-i_o)+j_o} y^{\beta(r-i_o)-j_o} z^{i_o} = x^{(\alpha+\beta)\tau^l - \tau^{l'}} y^{l'} z^{r-\tau^l}$$

is nonnull. It follows that the point $(\alpha+\beta-1/\tau^{l-l'}, 1/\tau^{l-l'})$ is in $\Delta(\mathcal{D}, E, p_2)$. Now, there exist a vertex in the segment joining (α,β) and $(\alpha+\beta, 0)$ of ordinate t with $0 < t < 1$, thus well prepared and then

(4.2.5.13) $$\varepsilon(\mathcal{D}, E, p_2) = 1.$$

(notice that $\beta \geq 1$).

(4.2.6) Remark. The third case is very much easier if the characteristic is zero. In fact, the possibility of eliminate the vertex $(\alpha+\beta, 0)$ created by $y \mapsto y_1$ in the hypersurface case is a phenomena typical of the positive characteristic.

(4.2.7) Theorem. Let $e(E) = 1$, let $p = (x,y,z)$ be a very well prepared normalized system of regular parameters, let $(X', E', \mathcal{D}', P')$ be a directional transform given from p by $(T-1,0)$ or T-3 and let $p' = (x',y',z')$ be the resulting system of parameters. Assume that if the transform is quadratic nor (x,z) nor (y,z) is permissible. Then one of the following properties is satisfied

a) A has won (as in 4.1.)

b) $\beta(\mathcal{D}', E', ') < \beta(\mathcal{D}, E, p)$.

c) $\beta(\mathcal{D}', E', p') = \beta(\mathcal{D}, E, p)$ and p' is a very well prepared normalized system of regular parameters.

Proof. Let us suppose that a) and b) are not satisfied. Then $\beta(\mathcal{D}',E',p') = \beta(\mathcal{D},E,p)$ by looking at (3.1.9.3). Moreover necessarily $\varepsilon(\mathcal{D},E,p) > 1$, and

(4.2.7.1) $$\varepsilon' = \varepsilon(\mathcal{D}',E',p') = \varepsilon(\mathcal{D},E,p)-1 = \varepsilon-1.$$

(this is not true if $\beta(\mathcal{D},E,p) \neq \beta(\mathcal{D}',E',p')$). By (4.1.4) p' is well prepared, so it is enough to prove that no coordinate change $y'_1 = y' + \zeta x'^{\varepsilon'}$ may increase $(\varepsilon',-1')$ if it is followed by a good preparation. Now reasoning as in (4.1.3) and (4.1.4) one can easily show that if $y'_1 = y' + \zeta x'^{\varepsilon'}$ followed by a good preparation may increase $(\varepsilon',-1')$, then

(4.2.7.2) $$y_1 = y + \zeta x^{\varepsilon} \qquad (\varepsilon = \varepsilon'+1)$$

followed by a good preparation may increase $(\varepsilon,-1)$ which is a contradiction.

(4.3) A winning strategy for type 0-1

(4.3.1) Here we shall put together the stability results above to obtain a winning strategy for the type 0-1. This strategy will depend on a first choice of a normalized well prepared system of parameters and of a construction in each step by the player A of a normalized well prepared system which depends (not in a unique way) of the preceding one and of the movement of the player B.

(4.3.2) (Winning strategy). Let $(X,E,\mathcal{D},P)$ be the status 0 of the type 0-1. Then, first of all, player A chooses a very well prepared normalized system of regular parameters $p(0) = (x(0),y(0),z(0))$, if $e(E) = 2$, A chooses $p(0)$ well prepared only. Now, we shall proceed by induction as follows:

Let $(X(t),E(t),\mathcal{D}(t),P(t))$ be the status t and let $p(t)$ be the ("very" if $e(E(t)) = 1$) well prepared normalized system of parameters which A has chosen. If $(x(t),z(t))$ or $(y(t),z(t))$ are permissible centers, then A chooses one of them for the center, otherwise A chooses the quadratic center. Now player B chooses a closed point $P(t+1)$ in the transform. If A does not win in this movement, then as an easy corollary of (4.1.2) the equation of the transformation is expressed from $p(t)$ in one of the following ways

(4.3.2.1) $(T\text{-}1,\zeta)$, T-2, T-3 or T-4.

Let $p'(t)$ be the obtained system of parameters. Then A applies a good preparation (as in (3.2.10) and (3.3.10)) to $p'(t)$ if $e(E(t+1)) = 2$ and a very good preparation (as in (3.4.6)) if $e(E(t+1)) = 1$. The result of this will be $p(t+1)$.

(4.3.3) Theorem. The strategy defined above is a winning strategy for the player A.

Proof. Let $G = \{G(t)\}$ be a realization of the game for which the above stratety has been applied. One has to prove that G is finite. Let us denote by $\{p(t)\}$ the sequence of systems of regular parameters used by A. We shall denote

(4.3.3.1)
$$(\beta(t),e(t),\varepsilon(t),\alpha(t)) = (\beta(\mathcal{D}(t),E(t),p(t)),e(E(t)),$$
$$\varepsilon(\mathcal{D}(t),E(t),p(t)),\ \alpha(\mathcal{D}(t),E(t),p(t))).$$

Since all the polygons $\Delta(\mathcal{D}(t),E(t),p(t))$ have their vertices placed in $(1/r!)\mathbb{Z}_0^{\,2}$ (easy to prove), in order to prove that G is finite it is enough to show that

(4.3.3.2)
$$(\beta(t),e(t),\varepsilon(t),\alpha(t)) < (\beta(t+1),e(t+1),\varepsilon(t+1),\alpha(t+1))$$

for the lexicographic order. If $\pi(t)$ is given by (T-1,0), T-2, T-3 or T-4, then prop. (3.1.9) proves (4.3.3.2) but putting $p'(t)$ instead of $p(t+1)$. We have always

(4.3.3.3)
$$\beta(t+1) = \beta(\mathcal{D}(t+1),E(t+1),p'(t))$$

in view of (4.1.3) and (4.1.4). Moreover, if $e(t) = e(t+1) = 2$ one has that

(4.3.3.4)
$$\alpha(t+1) = \alpha(\mathcal{D}(t+1),E(t+1),p'(t))$$
$$\varepsilon(t+1) = \varepsilon(\mathcal{D}(t+1),E(t+1),p'(t+1))$$

by theorem (4.1.3). If $e(t) = e(t+1) = 1$ and $\beta(t) = \beta(t+1)$ we have (4.3.3.4) by theorem (4.2.7). Then in these cases one has (4.3.3.2). Since $\beta(t+1) \leq \beta(t)$, we have (4.3.3.2) if $e(t) = e(t+1)$. If $e(t) = 1 < e(t+1) = 2$ one has (see (3.1.9)) that $\beta(t+1) < \beta(t)$. Now, it remains to consider the case of $(T\text{-}1,\zeta)$ with $\zeta \neq 0$. By theorems (4.2.5), (4.2.4) one has that (4.3.3.3) is true. Moreover, if $e(t) = 1$, then $e(t+1) = 1$ and $\beta(t+1) < \beta(t)$ in view of (4.2.5) c). If $e(t) = 2$, then $\beta(t+1) \leq \beta(t)$ but $e(t+1) = 1$. The proof is finished.

STANDARD TRANSITIONS FROM TYPE I

0. INTRODUCTION

(0.0.1) In this chapter, we shall continue the proof of the existence of a winning strategy for the reduction game in the case $n = 3$ and char $k = 0$. In view of the results if the chapter II, for the player A it is enough to reach a "type zero situation" in order to obtain the victory. Thus, we shall suppose that the game begins with a bit more general situation, called "type one". In this case, the technics of the polygon will not be as general as for the type zero. Actually, the biggest problem will be found in the control of "changes of situations", while the inner control in "each situation" will be provided by the polygon.

This chapter is mainly devoted to the study of the transitions which may be controled directly by the polygon, called "standard transitions". Also the first no standard transitions are treated.

1. CLASSIFICATION BY TRANSVERSALITY

(1.1) Ideals associated to a vector field

(1.1.1) The ideals which will serve us to define the transversality situations will be introduced here in a general manner.

(1.1.2) Let $(X,E,\mathcal{D},P)$ be such that $e(E) \geq 1$, $\dim \mathrm{Dir}\,(\mathcal{D},E) \geq 1$ and $r = \nu(\mathcal{D},E,P) \geq 2$.

(1.1.3) As in II(1.1.2), let us denote

(1.1.3.1) $$J_H^{\,r}(\mathcal{D},E) = J^r_{H(\mathrm{Dir}(\mathcal{D},E))}.$$

Let

(1.1.3.2) $$J^r(\mathcal{D},E) = \sum_{f \in R} J^r(D(f))$$

where D is a generator of $\mathcal{D}$.

Finally one has that

(1.1.3.3) $$J(\mathcal{D},E) = \begin{cases} J^r(\mathcal{D},E) & \text{if } J^r(\mathcal{D},E) \neq 0 \\ \sum_{I(E) \subset (f)} J^r(D(f)/f) & \text{if } J^r(\mathcal{D},E) = 0 \end{cases}$$

(1.1.4) The three ideals above are generated by linear forms in Gr(R), thus, one can consider

(1.1.4.1) $$J_H{*}(\mathcal{D},E) = J^r_H(\mathcal{D},E) \cap G_r^{\,1}(R)$$

and the same thing for $J^r{*}(\mathcal{D},E)$ and $J{*}(\mathcal{D},E)$, without loss of information. They are vector subspaces of $G_r^{\,1}(R)$, over the field k.

(1.1.5) Let us denote by J(E) the ideal of the directrix of E and let us denote by J C(E) the ideal of the tangent cone of E. As above, $J{*}(E) = J(E) \cap G_r^{\,1}(R)$.

(1.1.6) The ideal $\mathrm{In}^r(\mathcal{D},E)$ is defined by

(1.1.6.1) $$\mathrm{In}^r(\mathcal{D},E) = \sum_{f \in R} \mathrm{In}^r(D(f)).\mathrm{Gr}(R)$$

where D generates $\mathcal{D}$. One has that $In^r(\mathcal{D},E) = 0$ iff $\mu(\mathcal{D},E,P) = r$. The ideal $In(\mathcal{D},E)$ is defined as $In^r(\mathcal{D},E)$ if $\mu(\mathcal{D},E,P) = r-1$ and

(1.1.6.2) $$In(\mathcal{D},E) = \sum_{(f)\supset I(E)} In^r(D(f)/f).Gr(R)$$

where $\mu(\mathcal{D},E,P) = r$.

(1.2) Classification

(1.2.1) One has that

(1.2.1.1) $$J_H^r(\mathcal{D},E) \subset J^r(\mathcal{D},E) \subset J(\mathcal{D},E),$$

we shall give priority to the smallest of this ideals which is different from zero.

(1.2.2) Definition. $(X,E,\mathcal{D},P)$ is of the type "one" if $J^r{}_H(\mathcal{D},E) = 0 \neq J^r(\mathcal{D},E)$ and

(1.2.2.1) $$JC(E) \neq J^r(\mathcal{D},E) = J(\mathcal{D},E).$$

It is of the type "two" iff $J^r(\mathcal{D},E) = 0$,

(1.2.2.2) $$JC(E) \neq J(\mathcal{D},E)$$

and there is $\phi \in J^*(\mathcal{D},E)$ such that

(1.2.2.3) $$< \phi > + J^*(E) = Gr^1(R).$$

It is of the type "three" iff $J^r(\mathcal{D},E) = 0$, (1.2.2.2) is verified and there is no $\phi \in J^*(\mathcal{D},E)$ such that (1.2.2.3) is verified. Finally it is of the type "four" iff

(1.2.2.4) $$JC(E) = J_H(\mathcal{D},E)$$

or $$JC(E) = J^r(\mathcal{D},E) \text{ and } J_H(\mathcal{D},E) = 0$$

or $$JC(E) = J\ (\mathcal{D},E) \text{ and } J^r(\mathcal{D},E) = 0$$

(1.2.3) The conditions (1.2.2.1) and (1.2.2.2) mean that there is a component of E which is transversal to a form of $J(\mathcal{D},E)$.

(1.2.4) Definition. Let $(X,E,\mathcal{D},P)$ be of the type one. We shall say that it is of the type:

I-1 . If $\dim \mathrm{Dir}(\mathcal{D},E) = 1$ and $e(E) = 1$.

I'-1 . If $\dim \mathrm{Dir}(\mathcal{D},E) = 1$ and $e(E) = 2$.

I-2 . If $\dim \mathrm{Dir}(\mathcal{D},E) = 2$ and $e(E) = 1$.

I'-2-1 . If $\dim \mathrm{Dir}(\mathcal{D},E) = 2$, $e(E) = 2$, and $JC(E) \not\subset J^r(\mathcal{D},E)$.

I'-2-2 . If $\dim \mathrm{Dir}(\mathcal{D},E) = 2$, $e(E) = 2$ and $JC(E) \subset J^r(\mathcal{D},E)$.

(1.2.5) Lemma. If $(X,E,\mathcal{D},P)$ is of the type one, then it is of one of the types I-1,...,I'-2-2 above. Moreover, there is a regular system of parameters $p = (x,y,z)$ suited for (E,P) such that $\mathcal{D}$ is generated by

(1.2.5.1) $$D = ax\,\partial/\partial x + b\,\partial/\partial y + c\,\partial_z$$

where $\partial_z = \partial/\partial z$ or $\partial_z = z\partial/\partial z$ according to $e(E) = 1$ or $e(E) = 2$, in shuch a way that $\nu(c) \geq r+1$ if $e(E) = 1$, $\nu(b) = r$, (so $J^r(\mathcal{D},E) = J(\mathcal{D},E) = J^r(b)$) and

i) If $(X,E,\mathcal{D},P)$ is I-1, I'-1, then $J^r(b) = (\underline{x} = \underline{z} = 0)$.

ii) If $(X,E,\mathcal{D},P)$ is I-2, I'-2-2, then $J^r(b) = (\underline{z} = 0)$.

iii) If $(X,E,\mathcal{D},P)$ is I'-2-1, then $J^r(b) = (\underline{z} + \mu\underline{x})$, $\mu \neq 0$.

Proof. Let $e = 1$ and (x,y,z) be such that E is given by $x = 0$. If $\dim \mathrm{Dir}(\mathcal{D},E) = 2$, by (1.2.2.1) $J^r(\mathcal{D},E) \neq (\underline{x})$, so we can suppose that $J^r(\mathcal{D},E) = (\underline{z})$ and since $J^r_H = 0$, one has $\nu(c) \geq r+1$. If $\dim \mathrm{Dir}(\mathcal{D},E) = 1$, if $\underline{x} \notin J^r(\mathcal{D},E)$, then we can suppose $J^r(\mathcal{D},E) = (\underline{y},\underline{z})$ and this contradicts $J_H = 0$, so we can suppose $J^r(\mathcal{D},E) = (\underline{x},\underline{z})$, as above $\nu(c) \geq r+1$. Let $e = 2$ and let (x,y,z) be such that E is given by $xz = 0$. Necessarily $J^r(\mathcal{D},E) \subset J(E)$ since otherwise $J^r_H \neq 0$, so $J^r(\mathcal{D},E) = (\underline{z})$, $(\underline{z} + \mu\underline{x})$ or $(\underline{x},\underline{z})$ up to a change of order in x,z.

(1.2.6) Definition. Let $(X,E,\mathcal{D},P)$ be of the type I-1 or I'-1. $(X,E,\mathcal{D},P)$ is of the type I-1-0 (resp. I'-1-0) iff

(1.2.6.1) $$\mathrm{In}(I(E)) \not\supset \mathrm{In}^r(\mathcal{D},E)$$

and of the type I-1-1, resp. I'-1-1 otherwise. (In(I(E)) is the initial ideal of the ideal I(E) of E).

(1.2.7) <u>Remark</u>. If p = (x,y,z) is as in (1.2.5) and (X,E,$\mathcal{D}$,P) is of the type I-1-1 (resp. I'-1-1) then $\underline{x}$ divides $In^r(b)$ (resp. $\underline{x}\ \underline{z}$ divides $In^r(b)$).

(1.2.8) <u>Definition</u>. Let (X,E,$\mathcal{D}$,P) be of the type two. It is of the type

II-1-1 . If dim Dir ($\mathcal{D}$,E) = 1, e(E) = 2 and JC(E) $\not\subset$ J($\mathcal{D}$,E).

II'-1-1 . dim Dir ($\mathcal{D}$,E) = 1, E(E) = 3 and JC(E) $\not\subset$ J($\mathcal{D}$,E).

II-1-2 . dim Dir ($\mathcal{D}$,E) = 1, e(E) = 2 and JC(E) $\subset$ J($\mathcal{D}$,E).

II'-1-2 . dim Dir ($\mathcal{D}$,E) = 1, e(E) = 3, JC(E) $\subset$ J($\mathcal{D}$,E) and there is E' $\subset$ E n.c. div such that e(E') = 2 and JC(E') $\not\subset$ J($\mathcal{D}$,E).

II'-1-3 . dim Dir ($\mathcal{D}$,E) = 1, e(E) = 3, JC(E) $\subset$ J($\mathcal{D}$,E) and no II'-1-2.

II-2 . dim Dir ($\mathcal{D}$,E) = 2, e(E) = 2.

II'-2-1 . dim Dir ($\mathcal{D}$,E) = 2, e(E) = 3 and for each E' $\subset$ E n.c. div such that e(E') = 2 one has that J(E') + J($\mathcal{D}$,E) = Gr(R).

II'-2-2 . dim Dir ($\mathcal{D}$,E) = 2, e(E) = 3, JC(E) $\not\subset$ J($\mathcal{D}$,E) and there is E' $\subset$ E n.c. div. such that e(E') = 2 such that

$$J(E') + J(\mathcal{D},E) = Gr(R)$$

II'-2-3 . dim Dir ($\mathcal{D}$,E) = 2, e(E) = 3, JC(E) $\subset$ J($\mathcal{D}$,E).

(1.2.9) If we mark with a pointed contour the directrix, and with a continuous line the components of E, one can represent the above classification by the following figures:

II-1-1:

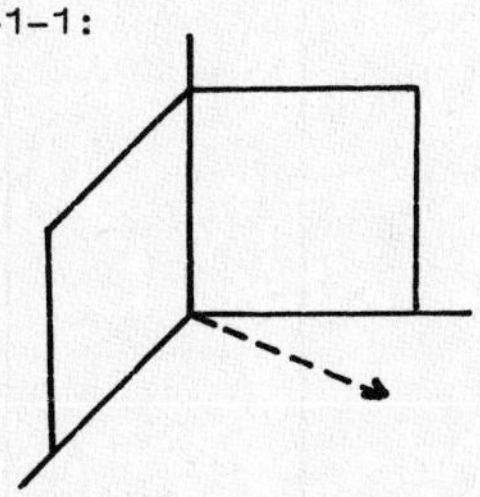

; II'-1-1:

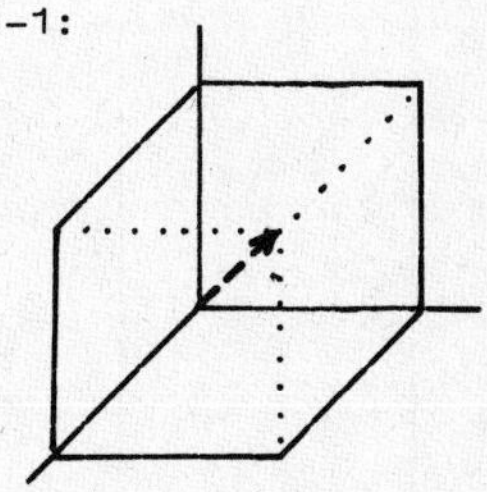

II-1-2:

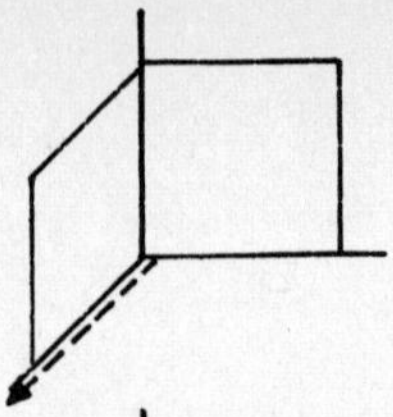

; II'-1-2:

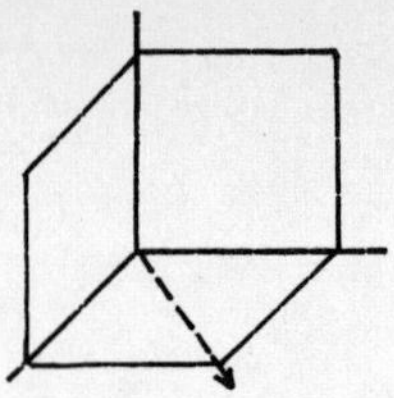

II'-1-3:

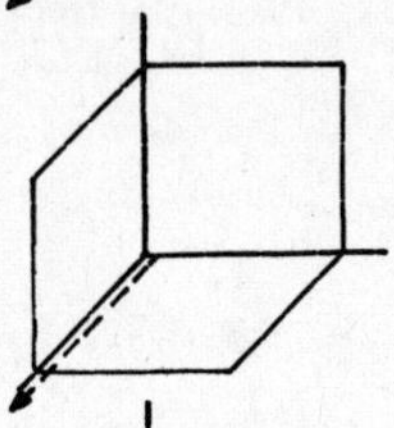

; II-2:

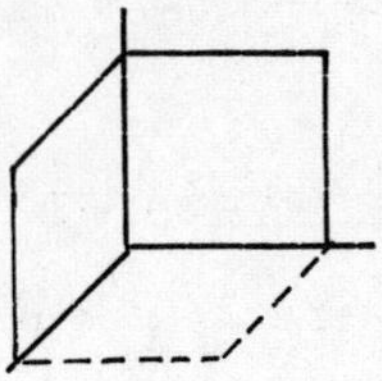

II'-2-1:

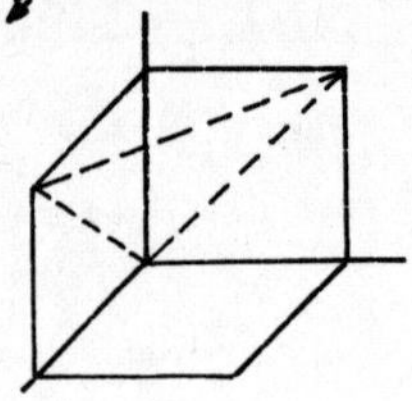

; II'-2-2:

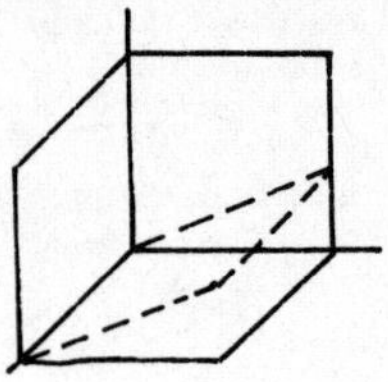

II'-2-3:

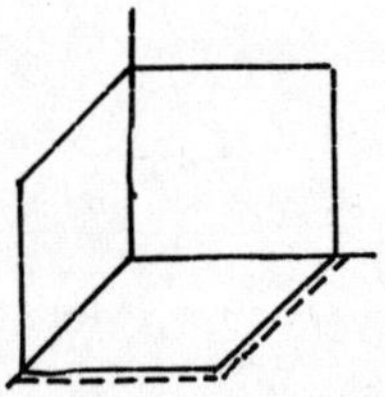

(1.2.10) Lemma. If $(X,E,\mathcal{D},P)$ is of the type two, then it is one of the types II-1-1,...,II'-2-3 above. Moreover, there is a regular system of parameters $p = (x,y,z)$ suited for (E,P) such that $\mathcal{D}$ is generated by

$$D = ax\partial/\partial x + y\partial/\partial y + c\partial_z \qquad (1.2.10.1)$$

where $\partial_z = \partial/\partial z$ or $z\partial/\partial z$ according to $e(E) = 2$ or $e(E) = 3$, such that $\nu(c) \geq r+1$ if $e(E) = 2$ and

i) If I-2-1, II'-1-2, then $J(\mathcal{D},E) = (\underline{y}+\lambda\underline{x}, z)$, $\lambda \neq 0$.

ii) If II'-1-1, then $J(\mathcal{D},E) = (\underline{y}+\lambda\underline{x}, \underline{z}+\mu\underline{y})$, $\lambda \neq 0$, $\mu \neq 0$.

iii) If II-1-2, II'-1-3, then $J(\mathcal{D},E) = (\underline{y},\underline{z})$.

iv) If II-2, II'-2-3, then $J(\mathcal{D},E) = (\underline{z})$.

v) If II'-2-1, $J(\mathcal{D},E) = (\underline{z}+\lambda\underline{x}+\mu\underline{y})$, $\lambda \neq 0 \neq \mu$.

vi) If II'-2-2, $J(\mathcal{D},E) = (\underline{z}+\lambda\underline{y})$, $\lambda \neq 0$.

Proof. If e = 1, clearly (1.2.5.1) is not possible, so $e(E) \geq 2$. Let us suppose $e(E) = 2$ and let (x,y,z) be a r.s. of p. such that E is given by $xy = 0$. Then $\nu(c) \geq r+1$ since $J^r = 0$. If dim Dir $(\mathcal{D},E) = 2$, then $J(\mathcal{D},E) = (\phi)$ with $\phi \notin J(E) = (\underline{x},\underline{y})$, so we can suppose that $\phi = \underline{z}$ and we have II-2. If dim Dir $(\mathcal{D},E) = 1$, as above, we can suppose $J(\mathcal{D},E) = (\alpha\underline{x}+\beta\underline{y},\underline{z})$ and we have II-1-1, II-1-2 according to $\alpha\beta \neq 0$ or $\alpha\beta = 0$.

Assume that $e(E) = 3$ and E is given by $xyz = 0$, then (1.2.2.3) is trivial and up to a change of order in the coordinates, we can suppose that $J(\mathcal{D},E) = (\underline{z}+\alpha\underline{x}+\beta\underline{y})$, or $(\underline{y}+\lambda\underline{x},\underline{z}+\mu\underline{x})$ and the result follows easily.

(1.2.11) Definition. Let $(X,E,\mathcal{D},P)$ be of the type II-1 (i.e. II-1-1 or II-1-2). Then $(X,E,\mathcal{D},P)$ is of the type II-1-1-0, resp. II-1-2-0, iff

(1.2.11.1) $$J(E) \not\supset In(\mathcal{D},E)$$

and of the type II-1-1-1, resp. II-1-2-1, otherwise.

(1.2.12) Remark. In the case $e(E) = 3$ there are more useful possibilities for dividing the types which will be treated in chapter V. By example, if we have type II'-2-3, we shall distinguish between the two cases:

i) If $\nu(f) = 1$, $I(E) \subset (f)$ and $In(f) \notin J(\mathcal{D},E)$ then $\nu(D(f)/f) \geq r+1$.

ii) $\exists f$, $\nu(f) = 1$, $I(E) \subset (f)$, $In(f) \notin J(\mathcal{D},E)$ such that $\nu(D(f)/f) = r$.

The first one will correspond to a more tangential situation than the second one.

(1.2.13) Definition. Let $(X,E,\mathcal{D},P)$ be of the type 3, we shall say that P is of the type

III-1-1 . If dim Dir $(\mathcal{D},E) = 1$, $e(E) = 1$ and $JC(E) \not\subset J(\mathcal{D},E)$.

III-1-2 . If dim Dir $(\mathcal{D},E) = 1$, $e(E) = 1$ and $JC(E) \subset J(\mathcal{D},E)$.

III'-1 . If dim Dir $(\mathcal{D},E) = 1$, $e(E) = 2$.

III-2 . If dim Dir $(\mathcal{D},E) = 2$, $e(E) = 1$.

III'-2-1 . If dim Dir $(\mathcal{D},E) = 2$, $e(E) = 2$ and $JC(E) \not\subset J(\mathcal{D},E)$.

III'-2-2 . If dim Dir $(\mathcal{D},E) = 2$, $e(E) = 2$ and $JC(E) \subset J(\mathcal{D},E)$.

(1.2.14) As in (1.2.8), one has the following pictures:

III-1-1: 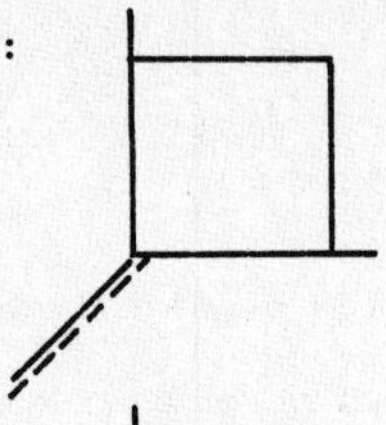 ; III-1-2: 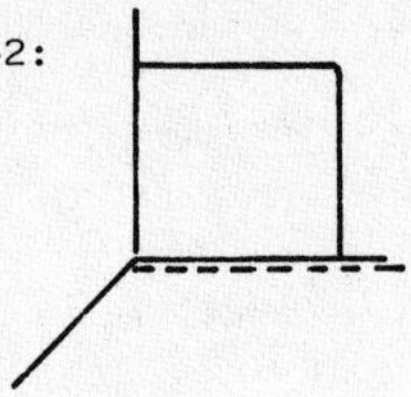

III'-1: 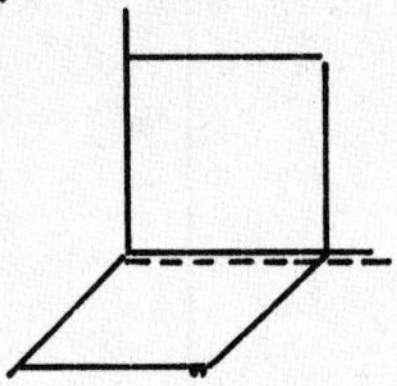 ; III-2: 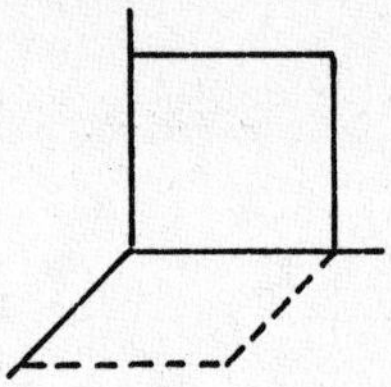

III'-2-1: 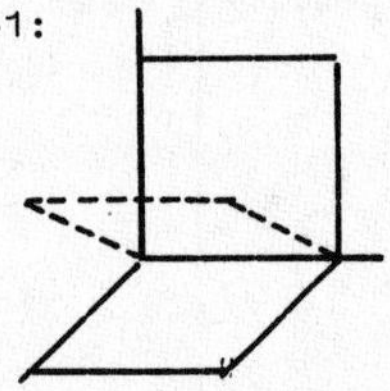 ; III'-2-2: 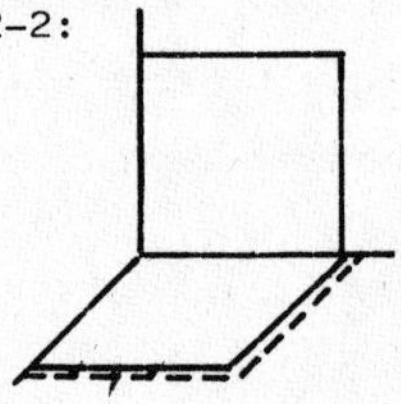

(1.2.15) Lemma. If $(X,E,\mathcal{D},P)$ is of the type 3, then it is of one of the types III-1-1,...,III'-2-2 above. Moreover, there is a r.s. of p. $p = (x,y,z)$ suited for (E,P) such that $\mathcal{D}$ is generated by

(1.2.15.1) $$D = ax\partial/\partial x + b\partial/\partial y + c\partial_z$$

where $\partial_z = \partial/\partial z$ or $z\partial/\partial z$ according to $e(E) = 1$ or 2, such that $\nu(b) \geq r+1$ and

i) If III-1-1, then $J(\mathcal{D},E) = (\underline{y},\underline{z})$.

ii) If III-1-2, III'-1 then $J(\mathcal{D},E) = (\underline{x},\underline{z})$.

iii) If III-2, III'-2-2 then $J(\mathcal{D},E) = (\underline{z})$.

iv) If III'-2-1, then $J(\mathcal{D},E) = (\underline{z}+\lambda\underline{x})$, $\lambda \neq 0$.

Proof. If $e = 1$, since $J^r = 0$ and we have type two, so transversality, one can choose (x,y,z) such that $x = 0$ gives E and $J(\mathcal{D},E) = (\underline{z})$, $(\underline{x},\underline{z})$ or $(\underline{y},\underline{z})$. Let us suppose $e = 2$ and take (x,y,z) such that E is given by $xz = 0$. If dim Dir $(\mathcal{D},E) =$

=1 since (1.2.2.3) is not verified, one has that $J(\mathcal{D},E) = (\underline{x},\underline{z})$. If dim Dir $(\mathcal{D},E) \neq 1$, necessarily $J(\mathcal{D},E) = (\lambda\underline{x}+\mu\underline{z})$ and, up to a change of order, one can suppose that $\mu \neq 0$.

(1.2.16) Definition. Let $(X,E,\mathcal{D},P)$ be of the type four, we shall say that it is of the type:

4-0, iff $J^r{}_H(\mathcal{D},E) \neq 0$.

4-1, iff $J_H(\mathcal{D},E) = 0 \neq J^r(\mathcal{D},E)$.

4-2, iff $J^r(\mathcal{D},E) = 0$.

(1.3) Reduction of the no transversals types

(1.3.1) Proposition. If $(X,E,\mathcal{D},P)$ is of the type four and $(X',E',\mathcal{D}',P')$ is a quadratic directional blowing-up such that $\nu(\mathcal{D}',E',P') = r$ and dim Dir $(\mathcal{D}',E',P') \geq 1$, then P' is of the type zero, one, two or three.

Proof. One has $e(E) = 1$ and the quadratic blowing-up must be made in a direction tangent to E, so $e(E') = 2$.

(1.3.2) Corollary. If there exists a winning strategy for the reduction game beginning at type zero, one, two or three, then there exists a winning strategy for the reduction game.

2. STANDARD TRANSITIONS

We shall begin with a situation I instead of 1 and, in the next chapter, we shall discuss the case in which the game begins with a situation 1. In this section, we shall describe the transition of "easy" control by means of the polygon as well as some results of reduction of the complexity of the possible transitions under the assumption of characteristic zero.

(2.1) Definitions and first reduction.

(2.1.1) Definition. Let $(X,E,\mathcal{D},P)$ be of type I,II or III and let $(X',E',\mathcal{D}',P')$ be a directional blowing-up of permissible center (may be quadratic) tangent to $\mathrm{Dir}\,(\mathcal{D},E)$. We shall say that $(X',E',\mathcal{D}',P')$ is a "standard" transition iff $r = \nu(\mathcal{D}',E',P')$ and one of the following possibilities is satisfied:

i) $(X,E,\mathcal{D},P)$ is of the type I and $(X',E',\mathcal{D}',P')$ is of the type I (we shall write I $\mapsto$ I).

ii) I $\mapsto$ II.

iii) II $\mapsto$ II.

iv) III $\mapsto$ II.

v) III $\mapsto$ III.

(2.1.2) Remark. The only no standard transitions between I,II,III are I $\mapsto$ III (which is impossible), II $\mapsto$ I and III $\mapsto$ I and, the transition III $\mapsto$ I has no interest for us since it will be considered as a "victory transition". Moreover, in this chapter we shall not be interested in the transtiions II $\mapsto$ III, III $\mapsto$ II and III $\mapsto$ III.

(2.1.3) Theorem. Let $G|_{s+1}$ be a partial realization of the reduction game such that for each $t = 1,\ldots,s$, $\pi(t)$ is a standard transition (see (I.(4.2)) for the notations). Assume that stat (0) is of the type one I and that the characteristic of k is zero. Then for each $t = 1,\ldots,s$, $\pi(t)$ is a transition I $\mapsto$ I, I $\mapsto$ II or II $\mapsto$ II.

Proof. One can assume that the first transition is I $\mapsto$ II. Let us suppose that stat (t) is of the type II for $t = 1,\ldots,s-1$ and that stat (s) is of the type III, in order to obtain a contradiction. Let us take $p = (x,y,z)$ as in (1.2.5). Then, the first transformation is given by T-2 or T-4 and $\mathcal{D}(1)$ is generated by

(2.1.3.1) $$a(1)x(1)\partial/\partial x(1) + b(1)y(1)\partial/\partial y(1) + c(1)\partial/\partial z(1)$$

(taking the notation of (1.2.5)) where $\nu(c(1)) \geq r+1$ and

(2.1.3.2)
$$a(1) = a/[y(1)]^{r-1}-b/[y(1)]^{r}$$
$$b(1) = b/[y(1)]^{r}$$
$$c(1) = c/[y(1)]^{r}-z(1)b/[y(1)]^{r}$$

if T-2, and

$$a(1) = a/[y(1)]^{r-1}$$
$$b(1) = b/[y(1)]^{r}$$
$$c(1) = c/[y(1)]^{r}-z(1)b/[y(1)]^{r}$$

ifT-4. We shall distinguish two cases: stat (0) is I-1 or it is I-2. If P is I-1, then T-4 has nosense, and if

(2.1.3.3)
$$\phi(\underline{x},\underline{z}) = \text{In }(b)$$

one has that

(2.1.3.4)
$$\text{In }(a(1)) = -\phi(\underline{x}(1),\underline{z}(1)) + \underline{y}(1)(\ldots)$$
$$\text{In }(b(1)) = \phi(\underline{x}(1),\underline{z}(1)) + \underline{y}(1)(\ldots).$$

Moreover, since dim Dir $(\mathcal{D}(1),E(1)) = 1$ (never = 2), after an adecuate change $z(1) \mapsto z(1)+\lambda.y(1)$ one can suppose that

(2.1.3.5)
$$\text{In }(a(1))= -\phi(\underline{x}(1) + \underline{y}(1),\underline{z}(1))$$
$$\text{In }(b(1))= +\phi(\underline{x}(1) + \underline{y}(1),\underline{z}(1)).$$

If $\mu= 0$, we continue by making T-2 until $t = s-1$. For $t = s-1$ we must have $J(\mathcal{D}(s-1,E(s-1)) = (\underline{x}(s-1)+\mu y(s-1),\underline{z}(s-1))$ where $\mu\neq 0$ since otherwise we cannont reach type III, and

(2.1.3.5)
$$\text{In }(a(s-1)) = -(s-1)\phi(x(s-1)+\mu\underline{y}(s-1),\underline{z}(s-1))$$
$$\text{In }(b(s-1)) = +\phi(\underline{x}(s-1)+ \underline{y}(s-1),\underline{z}(s-1))$$

Now we have to make $(T-1,1/\mu)$, but

(2.1.3.6)
$$\text{In }(b(s)) = (1/\mu)s.\phi(\underline{y}(s),\underline{z}(s))+x(s)(\ldots) \neq 0$$

(note $s \neq 0$ since the characteristic of k is zero) and thus it is not possible to have type III.

Assume that stat (0) is I-2. Then one can suppose that In (b) = $\underline{z}^r$. Then

(2.1.3.7)
$$\begin{aligned} \mathrm{In}\ (a(1)) &= \psi_1(\underline{x}(1),\underline{y}(1),\underline{z}(1)) \\ \mathrm{In}\ (b(1)) &= [\underline{z}(1)]^r + \phi_1(\underline{x}(1),\underline{y}(1),\underline{z}(1)) \end{aligned}$$

where $\psi_1(0,0,Z) = \phi_1(0,0,Z) = 0$. Moreover, after and adecuate change $z(1) \longmapsto z(1)+\lambda y(1)+\mu x(1)$ one can always suppose that $J(\mathcal{D}(1),E(1)) \ni \underline{z}(1)$. Moreover (x,y) is not permissible and then the next s-2 transformations are, following this procedure, of the type (T-1,0), T-2, T-3 or T-4 and for $t = s-1$, there exist $p,q \in \mathbb{Z}_0$ such that

(2.1.3.8)
$$\begin{aligned} \mathrm{In}\ (a(s-1)) &= -p[\underline{z}(s-1]^r+\psi_{s-1} \\ \mathrm{In}\ (b(s-1)) &= q[\underline{z}(s-1)]^r+\phi_{s-1} \end{aligned}$$

where $\psi_{s-1}(0,0,Z) = \phi_{s-1}(0,0,Z) = 0$. One can suppose that the following transformation is $(T-1,\zeta)$ and then

(2.1.3.8)
$$\mathrm{In}\ (b(s)) = \zeta.(p+q)[\underline{z}(s)]^r+\phi_s$$

with $\phi_s(0,0,Z) = 0$, then $\nu(b(s)) = r$, contradiction.

(2.1.4) Corollary. In a sequence $G|_{s+1}$ as in (2.1.3), the transition II $\mapsto$ III is not possible.

(2.2) Polygons and invariants

(2.2.1) Definition. Let $(X,E,\mathcal{D},P)$ be of the type I,II or III, a system of regular parameters $p = (x,y,z)$ is called a "normalized base" iff

a) I(E) = (x) or I(E) = (xy).

b) If one has type II-1-1-1 or II-1-2-1 or III-2, then $\underline{y} \notin J(\mathcal{D},E)$

If $(X,E,\mathcal{D},P)$ is of the type I', p is called a "normalized base" iff I(E) = (xz) and if E_2 is given by x and E_1 is given by z, then

(2.2.1.1) $$\mathrm{In}\,(I(E_1)) \not\ni \mathrm{In}^r(\mathcal{D},E) \implies \mathrm{In}\,(I(E_2)) \not\ni \mathrm{In}^r(\mathcal{D},E)$$

(a normalized base always exists).

(2.2.2) <u>Definition</u>. Let $(X,E,\mathcal{D},P)$ be of the type I,II or III and let $p = (x,y,z)$ be a normalized base. Let

(2.2.2.1) $$D = ax\partial/\partial x + b\partial_y + c\partial/\partial z$$

be a generator of $\mathcal{D}$. Then Exp(D,E;p) is defined by

(2.2.2.2)
$$\mathrm{Exp}\,(D,E,p) = \mathrm{Exp}\,(ya) \cup \mathrm{Exp}\,(b) \cup \mathrm{Exp}\,(yc/z)$$
$$\mathrm{Exp}\,(D,E,p) = \mathrm{Exp}\,(a) \cup \mathrm{Exp}\,(b) \cup \mathrm{Exp}\,(c/z)$$
$$\mathrm{Exp}\,(D,E,p) = \mathrm{Exp}\,(a) \cup \mathrm{Exp}\,(b/y) \cup \mathrm{Exp}\,(c/z)$$

for the types I,II and III respectively. $\mathrm{Exp}_+(D,E,p)$ is defined by

(2.2.2.3)
$$\mathrm{Exp}(D,E,p)\ (\text{for I}),\ \mathrm{Exp}(a)\ (\text{for III}) \text{ and}$$
$$\mathrm{Exp}(a) \cup \mathrm{Exp}(b)\ (\text{for II}).$$

The invariant $m\,(\mathcal{D},E,p)$ is the minimum h such that $(h,-1,r) \in \mathrm{Exp}\,(D,E,p)$ and $m(\mathcal{D},E,p) = \infty$ if such an h does not exist. (If $m \neq \infty$, then one has type III).

(2.2.3) <u>Definition</u>. Let $\psi:\{(h,i,j);\ j \leq r-1\} \to \mathbb{R}^2$ be given by $\psi(h,i,j) = (h/(r-j);i/(r-j))$. In the situation of (2.2.2), the polygon $\Delta(\mathcal{D},E,p)$ is defined by the convex hull of

(2.2.3.1)
$$[\psi(\mathrm{Exp}\,(D,E,p) \cap \{(h,i,j);j\leq r-1\}) +$$
$$+ \mathbb{H}(m(\mathcal{D},E,p))] \cap \{(u,v);v\geq -1\}$$

where $\mathbb{H}(m)$ is like in (II(2.2.7)). The polygon $\Delta_+(\mathcal{D},E,p)$ is defined by putting $\mathrm{Exp}_+(D,E,p)$ instead of Exp (D,E,p).

(2.2.4) <u>Remark</u>. In the above situation (x,z) is permissible iff $\Delta\,(\mathcal{D},E,p) \subset \{(u,v);\ u \geq 1\}$

and (y,z) is permissible iff $\Delta(\mathcal{D},E,p) \subset \{(u,v);\ v \geq 1\}$.

(2.2.5) Lemma. Let $(X,E,\mathcal{D},P)$ be of the types I,II or III and let $p = (x,y,z)$ be a normalized system of parameters. Let $(X',E',\mathcal{D}',P')$ be a directional blowing-up given by (T-1,0), T-2, T-3 or T-4 from p (T-3, resp. T-4, only if (x,z), resp. (y,z), is permissible). Moreover, let us usppose that it is a standard transition, then

(2.2.5.1) $$\Delta(\mathcal{D}',E',p') = \sigma(\Delta(\mathcal{D},E,p))$$

where σ is given as in (II.(3.1.9.4)) and p' is obtained from p by (T-1,0),T-2,T-3 or T-4.

Proof. If (T-1,0) or T-3 one has I $\mapsto$ I, II $\mapsto$ II or III $\mapsto$ III and if T-2 or T-4 one has I $\longmapsto$ II, II $\longmapsto$ II or III $\longmapsto$ II. A computation over the equations (I.(2.2.5)) gives the result. Let us remark that if III $\longmapsto$ III, (T-1,0) one has that $m' = m-1$.

(2.2.6) Definition. The invariants $\beta(\mathcal{D},E,p)$, $\varepsilon(\mathcal{D},E,p)$, $\alpha(\mathcal{D},E,p)$ are defined in the same way as in (II.(3.1.6)).

(2.2.7) Definition. Let $(X,E,\mathcal{D},P)$ be of the type I-1-1, II-1-1-1 or II-1-2-1 and let $p = (x,y,z)$ be a normalized base. The invariant $\delta(\mathcal{D},E,p)$ is defined by

(2.2.7.1) $$\delta(\mathcal{D},E,p) = \min\{ i/(r-h-j);\ (h,i,j) \in \mathrm{Exp}(D,E,p) \text{ and } h+j < r\}.$$

and $\delta_+(\mathcal{D},E,p)$ is defined by putting $\mathrm{Exp}_+(D,E,p)$ instead of Exp (D,E,p).

(2.2.8) Remarks. 1. $\delta = \infty$ if (x,z) is permissible.

2. $\Delta_+ \subset \Delta$, $\delta_+ \geq \delta$.

(2.3) Preparation of δ.

(2.3.1) Lemma. Let $(X,E,\mathcal{D},P)$ be of the type II-1-1-1 or II-1-2-1 and let $p=(x,y,z)$ be a normalized base. Let us consider the coordinate change

(2.3.1.1) $$z_1 = z + \sum_{n \geq \delta} \lambda_n y^n$$

where $\delta = \delta(\mathcal{D},E,p)$. Let $p_1 = (x,y,z_1)$. Then p_1 is normalized, one has that

(2.3.1.2) $$\delta(\mathcal{D},E,p_1) \geq \delta$$

and the equality in (2.3.1.2) occurs always if $\lambda_\delta = 0$.

Proof. Trivially p_1 is normalized. The monomials produced by (2.3.1.1) do not contribute in (2.2.7.1) to a point $t<\delta$ and if $\lambda_\delta = 0$, then the monomials which contribute to δ are not affected.

(2.3.2) Definition. Let $(X,E,\mathcal{D},P)$ be as above. A normalized base $p = (x,y,z)$ is "prepared" iff $\delta = \delta(\mathcal{D},E,p) \notin \mathbb{Z}_0$ or $\delta \in \mathbb{Z}_0$ and there is no change

(2.3.2.1) $$z_1 = z + \lambda y^\delta$$

such that $\delta(\mathcal{D},E,p_1) > \delta$ $(p_1 = (x,y,z_1))$. From a normalized base p one can always obtain a prepared base p' by making a (finite or not) sequence of coordinate changes as (2.3.2.1).

(2.3.3) Remark. The lemma (2.3.1) is true also if one put δ_+ instead of δ. This allows us to stablish the following:

(2.3.4) Definition. Let $(X,E,\mathcal{D},P)$ be of the type II-1-1-1 or II-1-2-1. A normalized base $p = (x,y,z)$ is "strongly prepared" iff it is prepared and $\delta_+ = \delta_+(\mathcal{D},E,p) \notin \mathbb{Z}_0$ or $\delta_+ \in \mathbb{Z}_0$ and there is no change

(2.3.4.1) $$z_1 = z + \lambda y^{\delta_+}$$

such that $\delta_+(\mathcal{D},E,p_1) > \delta_+$ $(p_1 = (x,y,z_1))$. From a prepared base p one can always obtain a strongly prepared base as in (2.3.2).

(2.3.5) Lemma. Let $(X,E,\mathcal{D},P)$ be of the type II-1-1-1 or II-1-2-1 and let $p=(x,y,z)$ be a strongly prepared base. Then

a) $J(\mathcal{D},E) = (\underline{x},\underline{z})$ if $\delta_+ > 1$.

b) $\delta_+ = 1$ iff $(X,E,\mathcal{D},P)$ is of the type II-1-1-1.

Proof. Since p is normalized: $J(\mathcal{D},E) = (\underline{x}+\lambda\underline{y},z+\mu\underline{y})$. One has $\delta_+ = 1$ iff $(\lambda,\mu) \neq 0$. If $\mu \neq 0$, $\lambda = 0$, then $z_1 = z + \mu y$ increases δ_+, contradiction. Thus $\mu \neq 0 \Rightarrow \lambda \neq 0$ and one has type II-1-1-1. Conversely, if type II-1-1-1, then $\lambda \neq 0$ and $\delta_+ = 1$.

(2.4) Standard transitions from the type I.1.1.

(2.4.1) Lemma. Let $(X,E,\mathcal{D},P)$ be of the type I-1-1 and let $(X',E',\mathcal{D}',P')$ be a quadratic directional blowing-up which corresponds to a standard transition. Then $(X',E',\mathcal{D}',P')$ is of the type II-1-1-1 or II-1-2-1.

Proof. Take $p = (x,y,z)$ as in (1.2.5). Necessarily the blowing-up is given by T-2. Then $\mathcal{D}'$ is generated as in (2.1.3.1) , (2.1.3.2). If $\nu(c') = r$, then $\underline{y}' | In(c')$ and the transition is not standard, thus $\nu(c') \geq r+1$ and in view of (2.1.3.4), (2.1.3.5) one has type II-1-1-1 or II-1-2-1.

(2.4.2) Theorem. Let $(X,E,\mathcal{D},P)$ be of the type II-1-2-1 and let $(X',E',\mathcal{D}',P')$ be a quadratic directional blowing-up which corresponds to a standard transition. Let $p = (x,y,z)$ be a strongly prepared base. Then

a) $(X',E',\mathcal{D}',P')$ is of the type II-1-1-1 or II-1-2-1.

b) The blowing-up is given by T-2 from Þ and if $p' = (x',y',z')$ is the resulting base, then p' is strongly prepared.

c) $\delta(\mathcal{D}',E',p') = \delta(\mathcal{D},E,p)-1$.

d) $\delta_+(\mathcal{D}',E',p') = \delta_+(\mathcal{D},E,p)-1$.

Proof. From 2.3.5. a), the blowing-up is given by T-2 and a), c) and d) follows straigthforward. If p' is not (strongly) prepared, the change $z'_1 = z'+\lambda y'^{\delta'}$ ($z'_1 = z'+\lambda y'^{\delta'_+}$) may be given back to a change $z_1 = z + \lambda y^{\delta}$ or $z_1 = z + \lambda y^{\delta+}$ and one obtains a contradiction.

(2.4.3) Corollary. Let G be a realization of the reduction game such that stat (0) is of the type I-1-1 and which follows the following strategy: if there is a permissible curve tangent to the directrix, then the player A chooses this center, otherwise he chooses the quadratic center. Assume that all the transitions in G are standard. Then G is finite.

Proof. By (2.4.1) and (2.4.2) the first transition is I-1-1 $\longmapsto$ II-1-2-1 (since if II-1-1-1 the following transition is not standard). The other transitions are II-1-2-1 $\longmapsto$ II-1-2-1 (one can assume that all the blowing-ups are quadratic, since otherwise the player A wins). Now it is enough to choose a strongly prepared base p(1) for stat (1) and to apply (2.4.2) c) (remark that $\delta<\infty$, since $\delta=\infty$ implies the existence of a permissible curve).

(2.5) Good preparation

(2.5.1) As in (II.(3.2)), we shall study the effect over the polygon of a change of the type

(2.5.1.1) $$z_1 = z + \xi x^{\alpha} y^{\beta}.$$

Since the computations will be quite similar to those in (II.(3.2)) and (II.(3.3)), we shall ommit or sketch most of them.

(2.5.2) Here $(X,E,\mathcal{D},P)$ will denote a situation of the types I-1-0, I-2, II-2, II-1-1-0 or II-1-2-0. Remark that with this assumption one has that

(2.5.2.1) $$(0,0,r) \in \operatorname{Exp}(D,E,p)$$

for each normalized regular system of parameters p.

(2.5.3) Lemma. Let us consider a set $A \subset \mathbb{Z}_0^2$ such that $A \subset \Delta(\mathcal{D},E,p)$, where $p=(x,y,z)$ is a normalized system of parameters. Let $A' \subset A$ be the set of vertices of $\Delta(\mathcal{D},E,p)$ which are in A. Let us consider the coordinate change

(2.5.3.1) $$z = z_1 + \sum_{(\alpha,\beta)\in A} \lambda_{\alpha\beta} x^\alpha y^\beta$$

Then one has that:

a) $p_1 = (x,y,z_1)$ is normalized.

b) $\Delta(\mathcal{D},E,p_1) \subset \Delta(\mathcal{D},E,p)$.

c) The vertices of $\Delta(\mathcal{D},E,p)$ not in A' are also vertices of $\Delta(\mathcal{D},E,p_1)$ and the monomials which contribute to them are exactly the same up to change of z by z_1.

Proof. This proof is similar to the proof of (II.3.2.3).

(2.5.4) Definition. Let p = (x,y,z) be a normalized system of parameters. A vertex $v = (\alpha,\beta)$ of $\Delta(\mathcal{D},E,p)$ is "well prepared" iff one of the following possibilities is verified:

a) $(\alpha,\beta) \notin \mathbb{Z}_0^2$.

b) There is no change of the type $z_1 = z + \lambda x^\alpha y^\beta$ (called "preparation of the vertex") such that $\Delta(\mathcal{D},E,p_1) \subset \Delta(\mathcal{D},E,p) - \{v\}$.

We shall say that $\Delta(\mathcal{D},E,p)$ (or p) is well-prepared iff all the vertices are well prepared

(2.5.5) Proposition. With notations as above, beginning with p = (x,y,z), one can reach a normalized system of parameters by a sequence of changes of vertex preparations, such that the corresponding polygon is well prepared.

Proof. It is a direct corollary of lemma (2.5.3).

(2.5.6) Remark. The lemma (2.5.3) is true also if one put $\Delta_+(\mathcal{D},E,p)$ instead of $\Delta(\mathcal{D},E,p)$. This allowsus to stablish the following:

(2.5.7) Definition. Let p = (x,y,z) be a normalized base and let (α,β) be a vertex of $\Delta(\mathcal{D},E,p)$ which is well prepared. The vertex (α,β) is said to be "strongly well prepared" iff one of the following possibilities is satisfied

a) $(\alpha,\beta) \notin \Delta_+(\mathcal{D},E,p)$

b) $(\alpha,\beta) \notin \mathbb{Z}^2$.

c) $(\alpha,\beta) \in \mathbb{Z}^2$, $(\alpha,\beta) \in \Delta_+(\mathcal{D},E,p)$ and there is no change of the type $z_1 = z+\lambda x^{\alpha} y^{\beta}$ such that $\Delta_+(\mathcal{D},E,p_1) \subset \Delta_+(\mathcal{D},E,p) - \{(\alpha,\beta)\}$. ($p_1=(x,y,z_1)$).

The base p is said to be "strongly well prepared" iff all the vertices of $\Delta(\mathcal{D},E,p)$ are strongly well prepared.

(2.5.8) One can obtain a strongly well prepared system of parameters from a normalized sistem p in the same way as in (2.5.5).

(2.6) Stability results for good preparation.

(2.6.1) Proposition. Let $(X,E,\mathcal{D},P)$ be as in (2.5.2). If $p = (x,y,z)$ is a strongly well prepared base, then $\underline{z} \in J(\mathcal{D},E)$.

Proof. If $\Delta_+(\mathcal{D},E,p) \cap \{u+v = 1\} = \emptyset$, then the result follows easily. Otherwise, if $\underline{z} + \lambda \underline{x} + \mu \underline{y} \in J(\mathcal{D},E)$ and vgr. $\lambda \neq 0$, the change $z_1 = z + \lambda x$ dissolves the vertex (1,0) of $\Delta_+(\mathcal{D},E,p)$, contradiction.

(2.6.2) Theorem. Let $(X,E,\mathcal{D},P)$ be as in (2.5.2), let $p = (x,y,z)$ be a strongly well prepared base, let $(X',E',\mathcal{D}',P')$ be a directional blowing-up which is quadratic or it is centered at (x,z), or (y,z), (the last cases only if the center is permissible). Assume that in the quadratic case it is given by T-1,0 or T-2 and that the transition is standard. Then

a) In the monoidal cases, the transformation is given by T-3 or T-4.

b) $(X',E',\mathcal{D}',P')$ is of the type I-2,I-1-0,II-2,II-1-1-0 or II-1-2-0.

c) If $p' = (x',y',z')$ is obtained from p by (T-1,0),T-2,T-3 or T-4, then p' is strongly well prepared.

Proof. a) follows from (2.6.1).

b) One has type I or II and computations as in the proof of (2.1.3) show that the fact for $In^r(\mathcal{D},E)$ of being or not in J(E) is preserved (remark that for type I, J(E) = In(I(E)).

c) Similar to (II.(4.1.3)).

(2.7) Very good preparation.

(2.7.1) As in (II.(3.4)) we are going to investigate the effect of a change $y_1 = y + \lambda x$ over the polygon, in order to have a control of the transformation $(T-1,\zeta)$ $\zeta \neq 0$.

The results of stability needed for the "controled situations" will be simpler that those of (II.(4.1)). We shall find results as in (II.(4.1)) in the study of transitions II $\longmapsto$ III, III $\longmapsto$ II, III $\mapsto$ III , II $\mapsto$ II that we shall make in chapter V.

Here $(X,E,\mathcal{D},P)$ will be of the type I-2 or of the type I-1-0. (but the last one would be irrelevant).

(2.7.2) Lemma. Let p = (x,y,z) be a strongly well prepared normalized system of parameters. Let us consider the coordinate change

(2.7.2.1) $$y_1 = y + \sum_{i \geq n} \lambda_i x^i .$$

Let $\Delta' = [\Delta(\mathcal{D},E,p) + \mathbb{H}(n)] \cap \mathbb{R}_o^2$. Then

a) $p_1 = (x,y_1,z)$ is normalized.

b) $\Delta'_1 = [\Delta(\mathcal{D},E,p_1) + \mathbb{H}(n)] \cap \mathbb{R}_o^2 = \Delta'$.

c) All vertices of $\Delta(\mathcal{D},E,p_1)$ which are vertices of Δ'_1 are strongly well prepared.

Proof. Similar to the proof of (II.(3.4.3)) (quite simpler).

(2.7.3) Definition. We shall say that p is "strongly very well prepared" iff it satisfies exactly the same conditions of (II.3.4.5), changing "well prepared" by "strongy well prepared".

(2.7.4) Proposition. There exist always a strongly very well prepared system of parameters for $(X,E,\mathcal{D},P)$ of type II-2.

Proof. See (II.(3.4.6)).

The proof of the following result is very similar to the proof of (II.(4.2.5)) (so it serves also for positive characteristic).

(2.7.5) Theorem. Let $(X,E,\mathcal{D},P)$ be of the type I-2 and $p = (x,y,z)$ be a strongly very well prepared base. Let us suppose that (x,z) does not define a permissible center. Let $y_1 = y + \zeta x$, $\zeta \neq 0$, and let $z \mapsto z_1$ be a strongly good preparation of $p_1 = (x,y_1,z)$. Let us denote $p_2 = (x,y_1,z_1)$. Then

a) The first vertex of $\Delta(\mathcal{D},E,p_2)$ is the same as the first vertex of $\Delta(\mathcal{D},E,p)$.

b) $\varepsilon(\mathcal{D},E,p_2) \leq 1$.

Proof. Similar to the proof of (II.(4.2.5)).

(2.8) Standard winning strategies.

(2.8.1) Here one obtains some winning strategies for the reduction game beginning at the type I-1-0 or I-2 and with the assumption that all the transitions are standard.

First of all, let us fix stat $(0) = (X,E,\mathcal{D},P)$ of the type I-1-0 or I-2 and let $p(0)$ be a strongly very well prepared base.

(2.8.2) Let G be a realization of the reduction game beginning at stat (0) and such that all the transitions are standard. In order to construct inductively regular systems of parameters $p(t)$, $t = 0,1,\ldots$, assume that G satisfies the following property: for each $t = 1,2,\ldots$, the center $Y(t-1)$ of $\pi(t)$ is the quadratic center or it is given by $(x(t-1),z(t-1))$ or $(y(t-1),z(t-1))$, where $p(t-1) = (x(t-1,y(t-1),z(t-1))$; moreover, $Y(t-1)$ is always permissible. Assume that $p(t-1)$ is strongly ("very" if stat $(t-1)$ is of the type I) well prepared. Then the above property implies that $\pi(t)$ is given by $(T\text{-}1,\zeta)$, T-2, T-3 or T-4. Let $p'(t)$ be the obtained regular system of parameters. Then $p(t)$ is defined to be a strong ("very" if stat (t) is of the type I) good preparation of $p'(t)$.

(2.8.3) Definition. Let G be a realization or the reduction game of length $\geq$ s+1, beginning at stat (0). G follows the "standard winning strategy" until the step s iff:

a) $\pi(t)$ is standard $t = 1,\dots,s$.

b) Let $p(t)$ be as above. If $(x(t),z(t))$ gives a permissible center, then $Y(t)$ is given by $(x(t),z(t))$. If $(x(t),z(t))$ is not permissible, but $(y(t),z(t))$ is permissible, then $Y(t)$ is given by $(y(t),z(t))$. $Y(t) = P(t)$ otherwise. $0 \leq t \leq s$.

(2.8.4) Definition. Let G be as above and let us fix $l \leq 0$. G follows the "l-retarded standard winning strategy" until the step s iff

a) $\pi(t)$ is standard $t = 1,\dots,s$.

b) If $(x(t),z(t))$ is permissible, then $Y(t)$ is given by $(x(t),z(t))$, $o \leq t \leq s$.

c) If $(x(t),z(t))$ is not permissible, $(y(t),z(t))$ is permissible, $\Delta(t) = \Delta(\mathcal{D}(t),E(t);p(t))$ has only one vertex and for each $t' < t$ such that $t-t' \leq l$ one has that $(y(t'),z(t'))$ is permissible, $\Delta(t')$ has only one vertex and $\pi(t'+1)$ is given by $(T\text{-}1,0)$ or T-3 from $p(t')$, then $Y(t)$ is given by $(y(t), z(t))$, $0 \leq t \leq s$.

d) If the assumptions of b) and c) are not verified, $(y(t),z(t))$ is permissible and $\Delta(t)$ has only one vertex, then $Y(t)$ is quadratic or $Y(t)$ is given by $(y(t),z(t))$. $0 \leq t \leq s$.

e) $Y(t)$ = quadratic center, otherwise. $0 \leq t \leq s$.

(2.8.5) Theorem. Let G be a realization of the reduction game beginning at stat (0) such that for a fixed strategy of (2.8.3) or (2.8.4), then G follows this strategy until s for each $s \leq$ length G. Then G is finite.

Proof. It follows from (2.2.5), (2.5.2) and (2.7.5). Remark that for the strategy of (2.8.3) and for the strategy 0-retarded, then $(\beta, \varepsilon, \alpha)$ decreases strictly each time. For the l-retarded case, may be (β, ∞, α) does not decrease, but β remains stable and after l-times, β decreases.

3. REDUCTION OF THE TYPE I-1-1.

In this section we shall consider all possible transitions and not merely the standard ones in order to prove the existence of a winning strategy for the reduction game when it begins with a situation I-1-1. We shall not prove directly the existence of a winning strategy: we shall prove that one of the two possibilities holds:

a) The player A wins.

b) The player A obtains a type I'-2 or I'-1-0.

In the following chapter, we shall prove that there exists a winning strategy for the reduction game beginning at type I'-2 or I'-1-0.

(3.1) No standard transitions.

(3.1.1) Here we shall define the special transitions different from the standard ones that we must to consider. In all the paragraph $(X,E,\mathcal{D},P)=(X(0),E(0),\mathcal{D}(0),P(0))$ will be of the type I-1-1, and $p = (x,y,z) = (x(0),y(0),z(0)) = p(0)$ a fixed strongly prepared system of parameters.

(3.1.2) We have $J(\mathcal{D},E) = (\underline{x},\underline{z})$. Thus in the first step of the game the player A will chose the center (x,z), if it is permissible, then the player A will always win. So, we can suppose that (x,z) is not permissible and that the player A choses the quadratic blowing-up.

(3.1.3) Let us suppose that $\mathcal{D}$ is generated by

(3.1.3.1) $$a(0)x(0)\partial/\partial x(0) + b(0)\partial/\partial y(0) + c(0)\partial/\partial z(0)$$

where necessarily $\nu(c(0)) \geq r+1$, $\nu(b(0)) = r$. If the player A does not win in the first movement, necessarily the player B must choose the transformation given by T-2. Thus we can suppose that $\mathcal{D}(1)$ is generated by

(3.1.3.2) $$a(1)x(1)\partial/\partial x(1) + b(1)y(1)\partial/\partial y(1) + c(1)\partial/\partial z(1)$$

where one has that

(3.1.3.3) $$\begin{aligned} a(1) &= a(0)/y(1)^{r-1} - b(0)/y(1)^{r} \\ b(1) &= b(0)/y(1)^{r} \\ c(1) &= c(0)/y(1)^{r} - z(1)b(0)/y(1)^{r} \end{aligned}$$

Now, there are two possibilities

(3.1.3.4) $$\nu(c(1)) \geq r+1 \quad \text{or} \quad \nu(c(1)) = r.$$

If $\nu(c(1)) \geq r+1$ and the player A has not won, then $(X(1),E(1),\mathcal{D}(1),P(1))$ is of the type I-1-1-1 or II-1-2-1 (see (2.4.2.)), and if $\nu(c(1)) = r$ and the player A has not won, one has type I', and the transition is not standard. We shall denote this last transition by

(3.1.3.5) $$I \longmapsto I'.$$

(3.1.4) Assume that in (3.1.3) one has had a standard transition. Let us denote by $p(1) = (x(1),y(1),z(1))$ a strongly prepared base for $(X(1),E(1),\mathcal{D}(1),P(1))$ (so we forget the notation of (3.1.3)). Then $\mathcal{D}(1)$ is generated as in (3.1.3.2), and $J(\mathcal{D}(1),E(1)) = (\underline{x}(1),\underline{z}(1))$ iff we have type II-1-2. If we have type II-1-1, by making if necessary a change $z_1 = z + \mu x$ we can assume $J(\mathcal{D}(1),E(1))=(\underline{y}(1)+\lambda\underline{x}(1),\underline{z}(1))$ where $\lambda \neq 0$ (see 2.3.5).

Let us suppose first that we have type II-1-2. Then if $(x(1),z(1))$ is permissible, the player A wins by choosing this center. So we can suppose that $(x(1),z(1))$ is not permissible and the player A chooses the quadratic blowing-up. Now, if the player A does not win in this movement, the player B must choose the transformation T-2 and one has that

$$a(2) = (a(1)-b(1))/y(2)^r$$

(3.1.4.1) $$b(2) = b(1)/y(2)^r$$

$$c(2) = c(1)/y(2)^{r+1}-z(2)b(1)/y(2)^r.$$

We have once more two possibilities

(3.1.4.2) $$\nu(c(2)) \geq r+1 \quad \text{or} \quad \nu(c(2)) = r.$$

If the player A has not won and $\nu(c(2)) \geq r+1$, then $(X(2),E(2),\mathcal{D}(2),P(2))$ is of the type II-1-2 or II-1-1 and $p(2) = (x(2),y(2),z(2))$ obtained from $p(1)$ by T-2 is a strongly prepared system of parameters (see (2.4.2)). If one has $\nu(c(2)) = r$, then one has that

(3.1.4.3) $$y(2) \mid \mathrm{In}\ (c(2))$$

and if the player A has not won, one has type I'. We shall denote this transition by

(3.1.4.3) $$\mathrm{II} \longmapsto \mathrm{I}'.$$

In this way, if the transitions $\mathrm{I} \mapsto \mathrm{I}'$ or $\mathrm{II} \mapsto \mathrm{I}'$ have not been produced, then after a finite number of steps we have a situation $(X(s),E(s),\mathcal{D}(s),P(s))$ of the type II-1-1 if A has not won, in view of th. (2.4.2). The nexttheorem assures that in this case the player A wins in the next movement.

(3.1.5) Theorem. With notations as above and if char (k) = 0, let us suppose that the player A has not won and $(X(t),E(t),\mathcal{D}(t),P(t))$, $t = 1,..,s-1$ are of type I-1-2 and for $t = s$ we have type I-1-1. Then if $(X(s+1),E(s+1),\mathcal{D}(s+1),P(s+1))$ is any directional quadratic transform of $(X(s),E(s),\mathcal{D}(s),P(s))$, one of the following possibilities is satisfied

a) $r > \nu(\mathcal{D}(s+1),E(s+1),P(s+1))$.

b) $\dim \mathrm{Dir}\ (\mathcal{D}(s+1),E(s+1)) = 0$.

c) $(X(s+1),E(s+1),\mathcal{D}(s+1),P(s+1)$ is of the type zero.

Proof. We shall suppose that a) and b) are not satisfied. Necessarily all the transformations are quadratic and intrinsically defined, so our reasonement does not depend on the particular choice of parameters as in (3.1.4) and (3.1.3). Actually, if there is a permissible curve tangent to the directrix in some step $t \geq 1$, (which is easily verified $t = 1$), for some strongly prepared system of parameters, one has $\varepsilon = \infty$ and it is not possible to reach by quadratic transformations as above the type I-1-1 (see (2.2.8) and (2.4.2)).

Let $p = (x,y,z) = p(0)$ be as in (3.1.1), one has that

(3.1.5.2) $$\mathrm{In}\,(b) = \phi(\underline{x},\underline{z})$$

where ϕ is not the power of a linear form. Now one can proceed as in theorem (2.1.3) just to obtain that for a certain $p(s+1) = (x(s+1),y(s+1),z(s+1))$ one has that with notation evident

(3.1.5.3) $$\mathrm{In}^r(b(s+1)) = \lambda.\,\phi(\underline{y}(s+1),\underline{z}(s+1)) + \underline{x}(\ldots)$$

and so, after an adecuate change, we can suppose that $\underline{y}(s+1),\underline{z}(s+1) \in J^r(b(s+1))$ and thus we have type zero.

(3.1.6) Remark. The winning strategy for A, if transitions I $\longmapsto$ I' and II $\longmapsto$ I' do not occur, is canonical and does not depend of the chosen coordinates for the control of the processus.

(3.1.7) So we have only to prove that there exists a winning strategy if in some step of the above processus (before reaching II-1-1) one has the transitions I $\longmapsto$ I' or II $\longmapsto$ I'.

(3.2) The transition I $\longmapsto$ I'. First cases.

(3.2.1) Let us take the notations of (3.1.3) and assume that after the first quadratic blowing-up given by T-2 from $p = (x,y,z)$, one has that

(3.2.1.1) $$\nu(c(1)) = r.$$

Now, in view of the equations (3.1.3.3) one has that

$$\underline{y}(1) \mid \mathrm{In}\,(c(1)). \tag{3.2.1.2}$$

and

$$\underline{y}(1) \in J^r(c(1)) = J(\mathcal{D}(1),E(1)). \tag{3.2.1.3}$$

In order to simplify the expressions of the future transformations, let us make the following change of coordinates

$$x'(1) = x(1);\ y'(1) = z(1);\ z'(1) = y(1) \tag{3.2.1.4}$$

Then $\mathcal{D}(1)$ is generated by

$$D = a'(1)x'(1)\partial/\partial x'(1) + b'(1)\partial/\partial y'(1) + c'(1)z'(1)\partial/\partial z'(1). \tag{3.2.1.5}$$

Where $a'(1) = a(1)$; $b'(1) = c(1)$; $c'(1) = b(1)$.

(3.2.2) One has that

$$\mathrm{In}\,(b) = \phi(\underline{x},\underline{z}) \tag{3.2.2.1}$$

where ϕ is not a power of a linear form. By (3.1.3.2) one has that

$$\begin{aligned} a'(1)(x'(1),y'(1),0) &= -\phi(x'(1),y'(1)). \\ b'(1)(x'(1),y'(1),0) &= -y'(1)\phi(x'(1),y'(1)). \\ c'(1)(x'(1),y'(1),0) &= \phi(x'(1),y'(1)). \end{aligned} \tag{3.2.2.2}$$

(we consider $a'(1),\ldots$ as series in $x'(1),y'(1),z'(1)$).

(3.2.3) Proposition. If there is a permissible curve tangent to Dir $(\mathcal{D}(1),E(1))$ for $(X(1),E(1),\mathcal{D}(1),P(1))$, then after the corresponding monoidal blowing-up, the adapted order drops in all the points.

Proof. Since $\underline{z}'(1) \in J(\mathcal{D}(1),E(1))$ and $(z'(1) = 0) \subset E(1)$, one has to consider curves $(z'(1),x'(1))$ or $(z'(1),y'(1)+\Phi(x'(1)))$ where $\nu(\Phi) \geq 1$. Assume first that $(x'(1),z'(1))$ gives a permissible curve. Then, necessarily

(3.2.3.1) $$\nu_{(x'(1),z'(1))}(a'(1)) \geq r$$

and this contradicts (3.2.2.2) since $\phi(x'(1),y'(1)) \neq x'(1)^r$. Thus this curve cannot be permissible.

Let us suppose that $(y'(1)+\phi(x'(1)),z'(1))$ is a permissible curve. Let us make the change

(3.2.3.2) $$x''(1) = x'(1);\ y''(1) = y'(1)+\phi(x'(1));\ z''(1) = z'(1)$$

and let us suppose that $\phi(x'(1)) = \lambda\, x'(1)+\ldots$. Then one has that

(3.2.3.3) $$a''(1)(x''(1),y''(1),0) = -\ \phi\,(\underline{x}''(1),\underline{y}''(1)-\lambda\underline{x}''(1))$$

Let us put

(3.2.3.4) $$\psi(x,y) = \phi(x,y-\lambda x).$$

Since $\nu_{(y)}(\psi) \geq r-1$, one has

(3.2.3.5) $$\psi\,(x,y) = \mu y^r + \sigma x y^{r-1}$$

where $\sigma \neq 0$. On the other way

(3.2.3.6) $$\mathrm{In}\ (b''(1)) = \underline{z}^r$$

(because one has type I' and the directrix being tangent to (y'',z'') cannot be given by $(\underline{x},\underline{z})$). Now the result follows from (3.2.3.5) and (3.2.3.6) since $r \geq 2$, by direct testing over the equations

(3.2.3.7) $$x''(1) = x''(2);\ y''(1) = y''(2);\ z''(1) = (z''(2)+\lambda)y''(2)$$
$$x''(1) = x''(2);\ y''(1) = y''(2)z''(2);\ z''(2) = z''(1).$$

(3.2.4) Remark. In view fo the above result, one can suppose that there is no permissible curves in $\mathcal{D}(1)$ tangents to the directrix in order to prove the existence of a winning strategy. Thus we shall assume that the player A chooses the quadratic center. Then, the equations of the transformation are necessarily given from $p'(1) = (x'(1),y'(1),z'(1))$ by $(T\text{-}1,\zeta)$ or T-2 if player A does not win. The following theorem shows that T-2 is not a good choice for the player B.

(3.2.5) Theorem. With notations as above, assume that $(X(2),E(2),\mathcal{D}(2),P(2))$ is a directional quadratic blowing-up given by T-2 from $p'(1) = (x'(1),y'(1),z'(1))$. Then one of the following possibilities is satisfied:

a) $r > \nu(\mathcal{D}(2),E(2),P(2))$.

b) $\dim \operatorname{Dir}(\mathcal{D}(2),E(2),P(2)) = 0$.

Proof. Assume that a) is not satisfied. Then $\mathcal{D}(2)$ is generated by

(3.2.5.1) $$D(2) = a'(2)x'(2)\partial/\partial x'(2) + b'(2)y'(2)\partial/\partial y'(2) + c'(2)z'(2)\partial/\partial z'(2)$$

where

(3.2.5.2) $$a'(2) = a'(1)/y'(2)^{r-1} - b'(1)/y'(2)^{r}$$ $$b'(2) = b'(1)/y'(2)^{r}$$ $$c'(2) = c'(1)/y'(2)^{r-1} - b'(1)/y'(2)^{r}.$$

In view of (3.2.2.2) one has

(3.2.5.3) $$a'(2)(x'(2),y'(2),0) = 0$$ $$b'(2)(x'(2),y'(2),0) = -y'(2)\phi(x'(2),1)$$ $$c'(2)(x'(2),y'(2),0) = 2.y'(2)\phi(x'(2),1)$$

Now, since the initial form of $b'(1)$ is

(3.2.5.4) $$\operatorname{In}(b'(1)) = \underline{z}'(1)\,.\,f(\underline{x}'(1),\underline{z}'(1))$$

one has that

(3.2.5.5) $$\operatorname{In}(a'(2)) = -\underline{z}'(1).f(\underline{x}'(1),\underline{z}'(1)) + \underline{z}'(1)\underline{y}'(1)(\ldots)$$

(see (3.2.5.3)). So

(3.2.5.6) $$\underline{z}'(1) \in J^{r}(a'(2)).$$

In the other hand, since ϕ is not a power of a linear form, and the order has not dropped, one has that

$$\phi(x'(2).1) = \gamma x'(2)^{r-1} + \delta x'(2)^{r} \tag{3.2.5.7}$$

where $\gamma \neq 0$. Then

$$\mathrm{In}\ (b'(2)) = \underline{z}'(2).f(\underline{x}'(2),\underline{z}'(2)) + \underline{z}'(2)\underline{y}'(2)(\ldots) + + \gamma\underline{y}'(2)\underline{x}'(2)^{r-1}. \tag{3.2.5.8}$$

Thus, $J^{r}(b'(2)) \neq (\underline{z} + \lambda\underline{x} + \mu\underline{y})$, since if this is true $\lambda \neq 0$ and $\underline{x}'(2)^{r}$ does not appear in the initial form. If dim Dir $(\mathcal{D}(2),E(2)) \geq 1$ one must have that

$$J^{r}(b'(2)) = (\underline{z} + \lambda\underline{x} + \mu\underline{y}\ ,\ \alpha\underline{x} + \beta\,\underline{y}) \tag{3.2.5.9}$$

Moreover, $\lambda\beta - \mu\alpha = 0$, since otherwise

$$J(\mathcal{D}(2),E(2)) \supset (\underline{z},\underline{z}+\lambda\underline{x}+\mu\underline{y},\alpha\underline{x}+\beta\underline{y}) = (\underline{z},\underline{x},\underline{y}) \tag{3.2.5.10}$$

So, we can suppose that

$$J^{r}(b'(2)) = (\underline{z}'(2),\ \alpha\underline{x}'(2) + \beta\,\underline{y}'(2)) \tag{3.2.5.11}$$

where $\alpha \neq 0$ in view of (3.2.5.8). This implies that

$$\mathrm{In}\ (b'(2)) = \Psi(\underline{z}'(2),\underline{x}'(2) + \beta\underline{y}'(2)) \tag{3.2.5.12}$$

Let us suppose that

$$\Psi(u,v) = \sum_{i+j=r} \mu_{ij}u^{i}v^{j}$$

then, in view of (3.2.5.8), necessarily $\mu_{or} \neq 0$ because the initial form is not divisible by $\underline{z}'(2)$. But in this situation, $\underline{x}'(2)^{r}$ must appear in the initial form: contradiction. Then

$$\dim \mathrm{Dir}\ (\mathcal{D}(2),E(2)) = 0. \tag{3.2.5.13}$$

(3.3) <u>The transition</u> I $\mapsto$ I'<u>. Case T-1 , ζ.</u>

(3.3.1) As we have seen in the precedent paragraph, it is enough to consider the case in which $(X(1),\mathcal{D}(1),E(1),P(1))$ has not permissible curves tangents to $\mathrm{Dir}\,(\mathcal{D}(1),E(1))$ and in the following quadratic transformation, the player B chooses one of the equations

$$(3.3.1.1)\qquad (T-1,\zeta)$$

form $p'(1) = (x'(1),y'(1),z'(1))$ (notation as in (3.2)).

(3.3.2) Let us introduce some notation, before starting to study this case. Let $p = (x,y,z)$ be a regular system of parameters, let us denote

$$(3.3.2.1)\qquad D^*(n;p) = -x\,\partial/\partial x + (n-1)y\partial/\partial y + (n+1)z\partial/\partial z$$

for $n = 0,1,2,\ldots$. Let us denote by $A^{**}(r;p)$ the subset of $\mathrm{Der}_k(R)\,|xz=0|$ composed by the vector fields

$$(3.3.2.2)\qquad D^{**} = ax\partial/\partial x + b\partial/\partial y + cz\,\partial/\partial z$$

such that $\nu(D^{**},\ xz = 0,\ P) = r-1$ and $\nu(b) = r-1$. The following lemma will simplify our task:

(3.3.3) Lemma. Let $p = (x,y,z)$ be a regular system of parameters of X at P and let E be given by $xz = 0$. Assume that $\mathcal{D}$ is generated by

$$(3.3.3.1)\qquad D = xy^{r-1}(1+\lambda x^{n-1}y)D^*(n;p) + x.D^{**} \qquad (n \geq 1).$$

where $D^{**} \in A^{**}(r;p)$. Let $\pi : X' \rightarrow X$ be a quadratic directional blowing-up and assume that the strict transform $(X',E',\mathcal{D}',P')$ satisfies that $r = \nu(\mathcal{D}',E',P')$ and $\dim \mathrm{Dir}\,(\mathcal{D}',E',P') \geq 1$. Then π is given by $(T-1,0)$ from p. Moreover, if $p' = (x',y',z')$ is obtained by $(T-1,0)$ from p, then $\mathcal{D}'$ is generated by

$$(3.3.3.2)\qquad D' = x'y'^{r-1}(1+\lambda x'^{(n+1)-1}y')D^*(n+1;p')+z'D'^{**}$$

where $D'^{**} \in A^{**}(r;p')$.

Proof. $z \in J(\mathcal{D},E)$. If π is given by T-2, then $\mathcal{D}'$ is generated by

$$(3.3.3.3)\qquad x'y'(1+\lambda x^{n-1}y^{n})(-nx'\partial/\partial x'+(n-1)y'\partial/\partial y'+2z'\partial/\partial z')+z'D''$$

If $r\geq 2$, $\nu(\mathcal{D}',E',P')\leq 2<r$. If $r=2$, a coefficient of D" has order 1 and dim $Dir(\mathcal{D},E')=0$. If π is given by $T-1,\zeta$, $\zeta\neq 0$, then $\mathcal{D}'$ is generated by

$$(3.3.3.4)\qquad x'(y'-\zeta)^{r-1}(1+x^n(y'-\zeta))(-x'\partial/\partial x'+n(y'-\zeta)\partial/\partial y'+(n+2)z'\partial/\partial z')+z'D''$$

and $\nu(\mathcal{D}'.E',P')\leq 1<r$. The second part of the lemma is trivial.

(3.3.4) Returning to the situation of (3.3.1), let us prove first that one can suppose $\zeta = 0$ without loss of generality. With notations as in (3.2), one has that $\mathcal{D}(1)$ is generated by

$$(3.3.4.1)\qquad D(1) = \phi(x'(1),y'(1))D^*(0,p'(1)) + z'(1)D^{**}$$

where $D^{**}\in A^{**}(r,p'(1))$.

Let $p'_\zeta(1) = (x'(1),y'_\zeta(1),z'(1))$, where $y'_\zeta(1) = y'(1) + \zeta x'(1)$. One has that

$$(3.3.4.2)\qquad D^*(0,p'(1)) = D^*(0,p'_\zeta(1))$$

$$(3.3.4.3)\qquad A^{**}(r,p'(1)) = A^{**}(r,p'_\zeta(1)).$$

And so, if $\phi_\zeta(x,y) = \phi(x,y+\zeta x)$ one has that

$$(3.3.4.4)\qquad D(1) = \phi_\zeta(x'(1),y'_\zeta(1))D^*(0,p'_\zeta(1)) + z'(1)D^{**}$$

where $D^{**}\in A^{**}(r,p'_\zeta(1))$. Since this decomposition is the only property that one needs, let us assume $\zeta = 0$.

(3.3.5) Let $p(2) = (x(2),y(2),z(2))$ be obtained from $p'(1)$ by $(T-1,0)$. Then $E(2)$ is given by $x(2).z(2) = 0$ and $\mathcal{D}(2)$ is clearly generated by

$$(3.3.5.1)\qquad D(2) = x(2)\phi(1,y(2))D^*(1,p(2)) + z(2)D^{**}(2).$$

If the player A has not won, necessarily one has that (except for a constant factor)

$$(3.3.5.2)\qquad \phi(x,y) = xy^{r-1} + \lambda y^r$$

(note that ϕ is not a power of a linear form).

Moreover, one has that

(3.3.5.3) $D^{**}(2) \in A^{**}(r,p(2))$.

So we can write

(3.3.5.4) $$D(2) = x(2)y(2)^{r-1}(1+\lambda x(2)^{-1}y(2))D^{*}(1,p(2))+z(2)D^{**}(2)$$

and we can apply the lemma (3.3.3), to obtain the following.

(3.3.6) Theorem. With notations as above, there are two possibilities:

i) $(y(2),z(2))$ defines a permissible curve for $(X(2),E(2),\mathcal{D}(2),P(2))$ and in this case if the player A chooses this center then he wins.

ii) $(y(2),z(2))$ does not define a permissible curve and in this case the player A wins in a finite number of steps by choosing always the quadratic center.

Proof. Let us take the notation of lemma (3.3.3). One has that (y,z) is a permissible curve for D iff (y',z') is a permissible curve for D'. Thus, by the results on stationnary sequences (I.(3.3)), if the player A does not won by choosing the quadratic center, then the player B has always chosen (T-1,0) by lemma (3.3.3) and then $(y(2),z(2))$ is a permissible curve. Now, one has to prove that if $\pi : X(3) \longrightarrow X(2)$ is a directional blowing-up with center $(y(2),z(2))$, then $\nu(\mathcal{D}(3),E(3),P(3)) < r$ or $\dim \mathrm{Dir}(\mathcal{D}(3),E(3)) = 0$. For this, let us observe that necessarily one has that

(3.3.6.1) $$\dim \mathrm{Dir}(\mathcal{D}(1),E(1)) = 2$$

and $J(\mathcal{D}(1),E(1)) = (\underline{z}'(1))$, since otherwise one would have type zero (or better) or the adapted order has dropped by making $(T-1,\zeta)$. Now, the result follows from (3.3.5.4) since $z(2)^r$ is a monomial which appears in the coefficient of $\partial/\partial y(2)$ in D(2).

(3.3.7) Remarks. 1. By (3.3.5.2), if $r > 2$, then there is only one point over P(1) after the quadratic blowing up which does not correspond to T-2 and the adapted order has not dropped. If $r = 2$ there is at most two such points.

2. Although now we are only interested in the reduction game be-

ginning at the type I-1-1, the above results are also valids for the type I-1-0 and so for the type I-1.

(3.4) The transition II $\mapsto$ I'.

(3.4.1) Assume now that one has a sequence

(3.4.1.1) $$(X(t),E(t),\mathcal{D}(t),P(t))$$

$t = 0,1,\ldots,s+1$, with $s \geq 1$, such that each step is a quadratic directional blowing--up of the precedent one and

(3.4.1.2) $$(X(0),E(0),\mathcal{D}(0),P(0)) = (X,E,\mathcal{D},P).$$

$$\nu(\mathcal{D}(t),E(t),P(t) = r \quad \forall t.$$

$$\dim \operatorname{Dir} (\mathcal{D}(t),E(t)) \geq 1 \quad \forall t.$$

If $t = 1,\ldots,s$ we have type II-1-2.

If $t = s+1$ we have type I'.

One has to prove that the player A can win also if the resolution game has had this feature. In this case the result will not be a direct way for the victory of the player A as in (3.3) and (3.2). The situation will be divided in two possibilities, the simpler one is analogous to the transition I $\mapsto$ I' and one can give an explicitid way for the player A to win. The other one will allow us to obtain a type I'-2 or I'-1-0.

(3.4.2) Remark. The sequence (3.4.1.2) is unique and it is the only choice for the player B. One can also suppose that there is no permissible curve tangent to the directrix in any step.

(3.4.3) Lemma. $p = (x,y,z) = p(0)$ may be chosen in such a way that for each $t=1,\ldots$ $\ldots,s+1$, the transformation

$$\pi(t)\colon X(t) \longrightarrow X(t-1)$$

is given by T-2 from $p(t-1) = (x(t-1), y(t-1), z(t-1))$ and $p(t) = (x(t), y(t), z(t))$ is the r.s. of p. obtained.

Proof. It is enough to make a strong preparation of $z(1)$ and "go back" to z in the usual way.

(3.4.5) Assume that $p = (x,y,z)$ has the above property, and that $\mathcal{D}(0)$ is generated by

$$(3.4.5.1) \qquad D(0) = ax\partial/\partial x + b\,\partial/\partial y + c\,\partial/\partial z$$

where $In(b) = \phi(\underline{x},\underline{z})$ and $\nu(c) \geq r+1$. Then $\mathcal{D}(1)$ is generated by

$$(3.4.5.2) \quad D(1) = \phi(x(1),z(1))(-x(1)\partial/\partial x(1) + y(1)\partial/\partial y(1) - z(1)\partial/\partial z(1)) + \\ + y(1)D^{**}(1)$$

where $D^{**}(1) \in Der_k(R(1)|E(1)|$. If we denote by

$$(3.4.5.3) \qquad D^{*\prime}(t;p) = (-tx\partial/\partial x + y\partial/\partial y - tz\partial/\partial z)$$

then $\mathcal{D}(s)$ is generated by

$$(3.4.5.4) \qquad D(s) = \phi(x(s),z(s))D^{*\prime}(x;p(s)) + y(s)D^{**}(s)$$

Since $J(\mathcal{D}(s),E(s)) = (\underline{x}(s),\underline{z}(s))$ by (3.4.3), one has that

$$(3.4.5.5) \qquad \nu(D^{**}(s)(x(s))/x(s)) \geq r; \quad \nu(D^{**}(s)(y(s))/y(s)) \geq r$$

Obviously: $\nu(D^{**}(s)(z(s))) \geq r$. Let us distinguish the two following posibilities:

$$(3.4.5.6) \qquad \nu(D^{**}(s)(z(s))) \geq r+1$$

$$(3.4.5.7) \qquad \nu(D^{**}(s)(z(s))) = r.$$

(3.4.6) Let us consider the first possibility (3.4.5.6). Then one has that $\mathcal{D}(s+1)$ is generated by

$$(3.4.6.1) \qquad D(s+1) = \phi(x(s+1),z(s+1))D^{*\prime}(s+1;p(s+1)) + y(s+1)D^{**}(s+1)$$

where if $p'(s+1) = (x'(s+1),y'(s+1),z'(s+1)) = (x(s+1),z(s+1),y(s+1))$ one has that $D^{**}(s+1) \in A^{**}(r,p'(s+1))$ (see 3.3.2). Now, for this case we can reason exactly as for the transition $I \longmapsto I'$ studied in (3.2) and (3.3). Actually, changing 1 by s+1 and 2 by s+2, prop. (3.2.3), th. (3.2.5) and th. (3.3.6) may be proved exactly in the same way. Then, in this case the player A wins in an explicit manner.

(3.4.7) Let us consider the possibility (3.4.5.7). Then $\mathcal{D}(s+1)$ is generated by

(3.4.7.1)
$$D(s+1) = \phi(x(s+1),z(s+1))D^{*\prime}(s+1;p(s+1)) + \\ + \psi(x(s+1),z(s+1)).\partial/\partial z(s+1) + y(s+1)D^{**}(s+1)$$

where ψ is homogeneus of degree r. If $\psi(x,z) \neq \psi(x,0)$, then one has type zero (or better) and so necessarily

(3.4.7.2)
$$\psi(x,z) = \lambda x^{r}, \quad \lambda \neq 0.$$

Now, if one makes a coordinate change $x = y(s+1)$, $y = z(s+1)$, $z = x(s+1)$, one deduces that $(X(s+1),E(s+1),\mathcal{D}(s+1),P(s+1))$ is of the type I'-2 or I'-1-0 (depending of $\nu(D^{**}(s+1),E(s+1)) = r-1$ or r). This proves the result in the beginning of this section.

(3.5) Reduction of the type I'-1-1.

(3.5.1) Theorem. If the reduction game begins at a situation of the type I'-1-1, then, by choosing the quadratic center if there is no permissible curve tangent to the directrix and such a curve if it exists, the player A wins or he obtains type I'-2 or I'-1-0.

Proof. Let $(X,E,\mathcal{D},P)$ be the status (0). Let $E = E_1 \cup E_2$ and let $p = (x,y,z)$ be such that $I(E) = (xz)$, $I(E_1) = (x)$, $I(E_2) = z$ If one is always in the strict transform of E_2, there is never a transition like $I \mapsto I'$ or $II \mapsto I'$ of (3.1) and the computations of (2.4) can be applied in order to obtain the victory.

Assume now that G is a realization of the game beginning at $(X,E,\mathcal{D},P)$ such that $P(s) \notin$ strict transform of E_2 (one can assume that $\pi(t)$ is quadratic for

$t \geq 2$). Then the above computations may be applied to the realization G' defined by

$$\text{stat}'(t) = (X(t), E'(t), \mathcal{D}(t), P(t))$$

where $E(t) = E'(t) \cup E_2(t)$ ($E_2(t)$ = strict transform of E_2). It is enough to remark that G' may be does not follow the winning strategy in the sense of not making the monoidal blowing-up in some step. But in this case, $\delta = \infty$, and then stat (s) is of the type II'-1-2 and there is a permissible center tangent to the directrix: the player A wins by choosing this center.

- IV -

A WINNING STRATEGY FOR TYPE ONE

0. INTRODUCTION

In this chapter it is continued the study of the reduction game when it begins with type one, in order to complete the proof of the existence of a winning stratety for this case.

All possible no standard transitions from the type I will be studied. In general, one cannot obtain directly the victory, but a special type, called "bridge type". But the study of the cases which give the victory is also useful for the study of the bridge type.

The end part of the chapter is devoted to the study of the reduction game beginning at the bridge type and also of the reduction game beginning at the type I'. In this way, the existence of a winning strategy for the reduction game beginning at the type one is proved.

In view of the results of the chapter III, the types I-1-1, or I'-1-1, are not considered as a beginning of the reduction game.

1. THE "NATURAL" TRANSITION

(1.1) Definition and notations

(1.1.1) In this section, $(X,E,\mathcal{D},P)$ will be of the type I (and, as we shall see, necessarily of the type I-2). Let us fix once for all a regular system of parameters $p = (x,y,z)$ such that it is strongly well prepared.

Moreover, assume that $x = z = 0$ is not a permissible curve for $(X,E,\mathcal{D},P)$. Let us fix $l \geq 0$ as "index of retardness".

(1.1.2) Definition. A "model for the natural transition" beginning at $(X,E,\mathcal{D},P)$, p is a sequence

(1.1.2.1) $$H = \{H(t) = (G(t),p(t))\}_{t=0,\dots,s+1}$$

with $s \geq 1$, such that:

a) $G|_{s+1}$ is a partial realization of the reduction game, beginning at $(X,E,\mathcal{D},P)$, such that $\pi(t)$, $t=1,\dots,s$ are standard, $e(E(1)) = 2$ (i.e. stat (1) is of the type II), and stat $(s+1)$ is of the type I.

b) $p(0) = p$, for each $t=1,\dots,s+1$, $p(t)$ is obtained from $p(t-1)$ by (T-1,0), T-2, T-3 or T-4. And $p(t)$ is strongly well prepared for $t = 1,\dots,s$.

c) $G|_{s+1}$ follows the l-retarded standard winning strategy relatively to $p(t)$ (see III (2.8.4)).

(1.1.3) Remarks. Before showing that the above model is useful for the control of certains transitions in the reduction game, let us make some evident remarks:

i) $\pi(1)$ is given by T-2 or T-4. If $\pi(1)$ is given by T-4, then $y = z = 0$ is a permissible curve for $(X,E,\mathcal{D},P)$ and $\Delta(\mathcal{D},E,p)$ has only one vertex.

ii) $\pi(s+1)$ is given by $(T-1,\zeta)$, $\zeta \neq 0$.

iii) $(X,E,\mathcal{D},P)$ must be of the type I-2, since as a consequence of (III.(3.1)), necessarily if we begin with type I-1, the no-standard transition given by $\pi(s+1)$ must to end in type I' (and not type I).

(1.1.4) Definition. Let $(X,E,\mathcal{D},P)$ be of the type II, and let $\pi: X' \to X$ be a directional quadratic blowing-up, and $(X',E',\mathcal{D}',P')$ the strict transform. We shall say that π is a "natural transition" iff

i) $r = \nu(\mathcal{D}',E',P') = \nu(\mathcal{D},E,P)$

ii) $(X',E',\mathcal{D}',P')$ is of the type I.

(1.1.5) Remark. Actually, we are interested in the first natural transition after a sequence of standard transitions begining at type I when the 1-retarded standard winning strategy has been applied. The main result in this section is that the final situation is better (strictly) than the initial one. The following proposition will allow us to work with a model.

(1.1.6) Proposition. Let G be a realization of the game beginning at $(X,E,\mathcal{D},P)$ and assume that the player A has followed in G the 1-retarded standard winning strategy with respect to p until the step s. Let us suppose too that stat (1) is of the type II and that $\pi(s+1)$ is a natural transition. Then, there exists $p' = (x',y',z')$ strongly well prepared such that

(1.1.6.1) $$\Delta(\mathcal{D},E,p') = \Delta(\mathcal{D},E,p)$$

and there exists a model H for the natural transition beginning at $(X,E,\mathcal{D},P),p'$ such that

(1.1.6.2) $$H(t) = (G(t),p'(t))$$

$t = 0,1,\ldots,s+1$.

Proof. It follows from (III. (2.6.1)) and (III. (2.6.2)).

(1.1.7) The rest of this section is devoted to prove the following theorem:

Theorem. Let H be a model for the natural transition beginning at $(X,E,\mathcal{D},P),p$.

Then there is a strongly well prepared system of regular parameters $p'(s+1)$ for $(X(s+1),E(s+1),\mathcal{D}(s+1),P(s+1))$ such that

(1.1.7.1) $$\beta(\mathcal{D}(s+1),E(s+1),p'(s+1)) < \beta(\mathcal{D},E,p).$$

(1.2) First reductions

(1.2.1) Lemma. Let $(X,E,\mathcal{D},P)$ be of the type I-2 (or I-1-0) and $p = (x,y,z)$ a normalized base such that $\mathcal{D}$ is generated by

(1.2.1.1) $$D = ax\partial/\partial x + b\partial/\partial y + c\partial/\partial z.$$

Assume that $(\alpha,\beta) \in \Delta(\mathcal{D},E,p)$ is a vertex, $(\alpha,\beta) \in \mathbb{Z}_0^{\,2}$ and that (α,β) is not well prepared. Then $(\alpha,\beta) \in \Delta^r(b;p)$ and it is not well prepared as a vertex of $\Delta^r(b;p)$.

(Remark . $\Delta^r(b;p)$ denotes the characteristic polygon

(1.2.1.2) $$\Delta^r(b;p) = \text{convex hull of } \{(h/(r-j);i/(r-j)); (h,i,j) \in \text{Exp}(b;p), j < r\} + \mathbb{R}_0^{\,2}).$$

Proof. A coordinate change $z_1 = z + \lambda x^\alpha y^\beta$ must eliminate (α,β) from $\Delta(\mathcal{D},E,p)$ Let $p_1 = (x,y,z_1)$ then

(1.2.1.3) $$D = a_1 x\partial/\partial x + b_1\partial/\partial y + c_1\partial/\partial z$$

where $a_1 = a$, $b_1 = b$, $c_1 = c + \lambda(\alpha x^\alpha y^\beta a + \beta x^\alpha y^{\beta-1} b)$. Necessarily $(\alpha,\beta) \notin \Delta^r(b;p')$, but since $(0,0,r) \in \text{Exp}(b;p)$, necessarily $(\alpha,\beta) \in \Delta^r(b;p)$ (by making the inverse change $z_1 \mapsto z$).

(1.2.2) Corollary. With notations as above, if $\Delta^r(b;p)$ is well prepared, then $\Delta(\mathcal{D},E,p)$ is well prepared.

(1.2.3) Lemma. Let $p = (x,y,z)$ be a r.s. of p. and $f \in R$, let us suppose that

(1.2.3.1) $$\{(h,i,r-1); (h,i) \in \mathbb{R}^2\} \cap \text{Exp}(f;p) = \emptyset.$$

Then $\Delta^r(f;p)$ is well prepared (characteristic zero!).

Proof. (It is the known "good property" of the Tchirnhausen transformation) If $(0,0,r) \notin \mathrm{Exp}\,(f;p)$ the result is trivial (the polygon is always well prepared). If $(0,0,r) \in \mathrm{Exp}\,(f;p)$ and (α,β) is a not well prepared vertex, then for each $j = 0,1,\ldots,r$ one has

$$(1.2.3.2) \qquad (\alpha j,\beta j,r-j) \in \mathrm{Exp}\,(f;p)$$

by considering the expansion of $(z+\lambda x^{\alpha}y^{\beta})^{r}$ (characteristic zero!).

(1.2.4) By the above lemmas one will obtain in an easy way the well prepared system of regular parameters $p'(s+1)$ of the theorem (1.1.7). The characteristic zero of the base field is importante here.

(1.2.5) Proposition. Let $H = \{G, p(t)\}$ be a model for the natural transition beginning at $(X,E,\mathcal{D},P),p$. Then there is another model $H' = \{G', p'(t)\}$ such that $G = G'$, $\Delta(\mathcal{D},E,p) = \Delta(\mathcal{D},E,p')$ and $p'(s+1)$ is well prepared.

Proof. Let us suppose that $\mathcal{D}(t)$ is generated by

$$(1.2.5.2) \qquad D(t) = a(t)x(t)\partial/\partial x(t)+b(t)y(t)\partial/\partial y(t)+c(t)\partial/\partial z(t)$$

with respect to $p(t) = (x(t),y(t),z(t))$, $t = 1,\ldots,s$. Since $\pi(1)$ is given by T-2 or T-4 and for $t = 2,\ldots,s$ $\pi(t)$ is given by (T-1,0), T-2, T-3 or T-4 one can deduce by induction that there exists $\lambda \neq 0$, $n(t) \geq 0$, $m(t) > 0$ such that

$$(1.2.5.3) \qquad \begin{aligned} a(t)(0,0,z) &= -\,n(t)z^{r} + \text{(higher degree terms)} \\ b(t)(0,0,z) &= m(t)z^{r} + \text{(higher degree terms)} \end{aligned}$$

Let $f = b(s)-a(s)$. Clearly $(0,0,r) \in \mathrm{Exp}\,(f;p(s))$. Let $\Delta = \Delta^{r}(f;p(s))$. One has $\Delta \subset \Delta(\mathcal{D}(s),E(s);p(s))$. After a change of coordinates

$$(1.2.5.4) \qquad z'(s) = z(s) + \sum_{(\alpha,\beta)\in\Delta} \lambda_{\alpha\beta}x(s)\ y(s)$$

one has that, if $p'(s) = (x(s),y(s),z'(s))$, then

$$(1.2.5.5) \qquad \{(h,i,r-1);(h,i) \in \mathbb{R}^{2}\} \cap \mathrm{Exp}\,(f;p'(s)) = \emptyset.$$

Since $\Delta \subset$ convex hull of $\Delta^r(a(s);p(s)) \cup \Delta^r(b(s);p(s))$, one deduces that p'(s) is strongly well prepared. The change (1.2.5.4) may be "given back" to $(X(s-1),E(s-1),\mathcal{D}(s-1),P(s-1))$ in the usual way, changing p(s-1) by p'(s-1) and by induction just to $(X(0),E(0),\mathcal{D}(0),P(0))$. Moreover, the polygon is not modified in each step. Let us observe that

(1.2.5.6) $$D(s) = a'(s)x(s)\partial/\partial x(s) + b'(s)y(s)\partial/\partial y(s) + c'(s)\partial/\partial z'(s)$$

where a'(s) = a(s), b'(s) = b(s). Now, if we make $T-1,\zeta$ from p'(s) to obtain (s+1), we have

(1.2.5.7)
$$D(s+1) = a'(s+1)x'(s+1)\partial/\partial x'(s+1)+b'(s+1)\partial/\partial y'(s+1)+c'(s+1)\partial/\partial z'(s+1)$$

Clearly p'(s+1) is normalized and, since

(1.2.5.8) $$b'(s+1) = (b(s)-a(s))(y'(s+1)+\zeta)/x'(s+1)^r$$

one has that

(1.2.5.9) $$\{(h,i,r-1)\} \cap \mathrm{Exp}\,(b'(s+1);p'(s+1)) = \emptyset$$

and by applying lemmas (1.2.1) and (1.2.3), p'(s+1) is well prepared.

(1.3) Proof of the main result

(1.3.1) Here the theorem (1.1.7) is proved. In view of the above reductions, it is enough to take a fixed model H such that p(s+1) is well prepared and to prove that

(1.3.1.1) $$\beta(\mathcal{D}(s+1),E(s+1),p(s+1)) < \beta(\mathcal{D},E,p).$$

Let us simplify the notation by writting

(1.3.1.2) $$\beta(t) = \beta(\mathcal{D}(t),E(t),p(t))$$
$$\Delta(t) = \Delta(\mathcal{D}(t),E(t),p(t)). \qquad t=0,1,...,s+1$$

(1.3.2) Proposition. $\beta(s) < \beta(0)$ and

(1.3.2.1) $$\Delta(s) \cap \{(u,0);\ u \in \mathbb{R}\} = \emptyset.$$

Proof. Since $\pi(1)$ is given by T-2 or T-4, one has that $\beta(1) < \beta(0)$. Then the first assertion is satisfied since $\beta(t) \leq \beta(t-1)$, $t = 1,\ldots s$.

Now, let us proceed by induction. If $\pi(1)$ is given by T-2, then $\Delta(1) = \sigma(\Delta(0))$, where $\sigma(u,v) = (u,u+v-1)$, then

(1.3.2.2) $$\Delta(1) \cap \{(u,0)\} \neq \emptyset$$

iff $(1,0) \in \Delta(0)$. But $(1,0) \notin \Delta(0)$ since $(X,E,\mathcal{D},P)$ is of the type I-2 and $p(0)$ is well prepared (Remark that this is not true in general for type II-2). If $\pi(1)$ is given by T-4, then $\Delta(0)$ must have only one vertex; $\Delta(1) = \sigma(\Delta(0))$ where $\sigma(u,v) = (u,v-1)$. Then one has (1.3.2.2) iff $(1,1)$ is the only vertex of $\Delta(0)$, but in this case we must to make T-3 instead of T-4. Let us suppose that

(1.3.2.3) $$\Delta(t) \cap \{(u,0)\} = \emptyset$$

where $1 \leq t \leq s$. If $\pi(t+1)$ is given by (T-1,0) or T-3, then $\Delta(t+1) \cap \{(u,0)\} = \emptyset$ clearly. If $\pi(t+1)$ is given by T-2 or T-4 we shall reason as above.

(1.3.3) $\pi(s+1)$ is given by $(T\text{-}1,\zeta)$, $\zeta \neq 0$, and the coefficients of a generator of $\mathcal{D}(s+1)$ are given by

(1.3.3.1) $$a(s+1) = a(s)/x(s+1)^r$$
$$b(s+1) = (y(s+1)-\zeta)(b(s)-a(s))/x(s+1)^r$$
$$c(s+1) = c(s)/x(s+1)^{r+1} - z(s+1)a(s)/x(s+1)^r$$

where $a(s),b(s),c(s)$ are the coefficients of a generator of $\mathcal{D}(s)$. Let us denote $y_1(s) = y(s) + \zeta x(s)$ and $p_1(s) = (x(s),y_1(s),z(s))$. Let $p_0(s) = p(s)$ and let us denote by "[[...]]" the "convex hull", let us introduce the auxiliary polygons

(1.3.3.2) $$\Delta^*(i) = [[\Delta^r(a(s);p_i(s)) \cup \Delta^r(b(s)-a(s);p_i(s)) \cup \Delta^{r+1}(c(s)-z(s)a(s);p_i(s))]], \qquad i=0,1$$

(1.3.3.3) $$\Delta^{*\prime}(1) = \sigma(\Delta^*(1)), \text{ where } \sigma(u,v) = (u+v-1,v)$$

Easy computations show that:

(1.3.3.4) $$\Delta^*(0) = \Delta(s)$$

$$\Delta^*(1) + \mathbb{H}(1) = \Delta(s) + \mathbb{H}(1)$$

(the last assertion follows from the effect of a change $y_1 = y+\lambda x$ in the polygon of a surface).

(1.3.4) Lemma. $\Delta^{*\prime}(1) + (0,1) \subset \Delta(s+1) \subset \Delta^{*\prime}(1)$. Moreover, the main vertices of $\Delta^{*\prime}(1)$ and $\Delta(s+1)$ have the same abscissa.

Proof. Let $(p',q') \in \Delta(s+1)$, since

(1.3.4.1) $$\Delta(s+1) = [[\Delta^r(y(s+1)a(s+1);p(s+1)) \cup \Delta^r((b(s)-a(s))/x(s+1)^r;p(s+1)) \cup \cup \Delta^{r+1}(y(s+1)[(c(s)-z(s)a(s)]/x(s+1)^{r+1};p(s+1)]].$$

then there exists $(p,q) \in \Delta(s+1)$ such that $(p',q') \in (p,q) + \mathbb{R}_0^2$ (if (p',q') is a vertex then $(p,q) = (p',q')$) and there exist $j < r$ such that

(1.3.4.2) $$(p(r-j),q(r-j),j) \in Exp(y(s+1)a(s+1);p(s+1)) \cup \cup Exp((b(s))/x(s+1)^r;p(s+1)) \cup \cup Exp(y(s+1)[c(s)-z(s)a(s)]/x(s+1)^{r+1}z(s+1);p(s+1)).$$

Now, looking at (1.3.3.2) and (1.3.3.3) one deduces that

(1.3.4.3) $$(p,q) \in \Delta^{*\prime}(1) \quad \text{or} \quad (p,q-1/(r-j)) \in \Delta^{*\prime}(1).$$

This proves the last assertion and $\Delta(s+1) \subset \Delta^{*\prime}(1)$. The proof of $\Delta^{*\prime}(1)+(0,1) \subset \Delta(s+1)$ is obtained by reversing the arguments.

(1.3.5) Let $\Delta \subset \mathbb{R}_0^2$ be a polygon. Let us define $\delta(\Delta)$ the number such that

(1.3.5.1) $$\dot{\Delta} \cap \{u+v = \delta(\Delta)\} = \emptyset$$

$$\partial(\Delta) \cap \{u+v = \delta(\Delta)\} \neq \emptyset.$$

$(\dot{\Delta} = \Delta - \partial(\Delta))$.

Let us denote by

(1.3.5.2) $$v^*(\Delta) = (\alpha^*(\Delta),\beta^*(\Delta)),$$

resp. $$v^{*\prime}(\Delta) = (\alpha^{*\prime}(\Delta),\beta^{*\prime}(\Delta))$$

the vertex of Δ of smallest (resp. biggest) abscissa in

(1.3.5.3) $$\partial(\Delta) \cap \{u+v = \delta(\Delta)\}$$

(Remark that it is possible to have $v^*(\Delta) = v^{*\prime}(\Delta)$).

(1.3.6) Lemma. Let $f \in k[[x,y,z]]$, $\nu(f) \geq r$. Let $\Delta = \Delta^r(f;(x,y,z))$ and let $y_1 = y+\zeta x$, $\zeta \neq 0$. Let $\Delta_1 = \Delta^r(f;(x,y_1,z))$. Then one has that

i) $v^*(\Delta) = v^*(\Delta_1)$

ii) $\beta^{*\prime}(\Delta_1) \leq \beta^*(\Delta) - \beta^{*\prime}(\Delta)$.

Proof. i) See (II. 3.4)

ii) Let us suppose that

(1.3.6.1) $$f = \sum f_{hij} x^h y^i z^j.$$

First, let us observe that we can suppose that

(1.3.6.2) $$f_{hij} \neq 0 \Rightarrow (h+i)/(r-j) = \delta(\Delta) \text{ and } j < r.$$

(we can forget the others monomials for our purposes). Now, working by induction on r and remarking that the power of x which divides f is not affected by $y \mapsto y_1$, we can suppose that

(1.3.6.3) $$f = \sum_{a \leq l \leq b} f_l x^{b-l} y^l, \quad f_b \neq 0$$

where $a = r.\beta^{*\prime}(\Delta)$, $b = r.\beta^*(\Delta)$. In this situation ii) is not satisfied iff for each $s = 0,1,\ldots,b-a$, one has

(1.3.6.4) $$\sum_{a \leq l \leq b} f_l \binom{l}{s} \zeta^{l-s} = 0.$$

Thus the above system has a no trivial solution in the indeterminates f_l, since $f_b \neq 0$. To obtain contradiction it is enough to prove that the rank of its matrix is precisely $b-a+1$ = number of indeterminates. The matrix of (1.3.6.4) is

(1.3.6.5) $$A = (a_{nm})_{o \leq n,m \leq b-a}$$

where

(1.3.6.6) $$a_{nm} = \binom{a+m}{n} \zeta^{a+m-n}$$

Then

(1.3.6.7) $$\det A = \zeta^{(b-a+1).a} \neq 0.$$

(1.3.7) In order to finish the proof of the theorem (1.1.9). Let us consider a fixed (h,i,j) such that

(1.3.7.1) $$h/(r-j) = \alpha^*(\Delta^*(0)), \quad i/(r-j) = \beta^*(\Delta^*(0))$$

(1.3.7.2) $$(h,i,j) \in \mathrm{Exp}\,(a(s);p_o(s)) \cup \mathrm{Exp}\,(b(s)-a(s);p_o(s)) \cup \cup \mathrm{Exp}\,([c(s)-z(s)a(s)]/z(s);\ p_o(s)).$$

Now, by the lemma (1.3.6) and by the proposition (1.3.2) (which implies that $\beta^{*\prime}(\Delta^*(0)) \neq 0$) one has that there exist $l > 0$, $l \in \mathbb{N}$ such that

(1.3.7.3) $$(h+l,i-l,j) \in \mathrm{Exp}(a(s);p_1(s)) \cup \mathrm{Exp}(b(s)-a(s);p_1(s)) \cup \cup \mathrm{Exp}([c(s)-z(s)a(s)]/z(s);\ p_1(s)).$$

This implies that

(1.3.7.4) $$(h+i+j-r,i-l,j) \in \mathrm{Exp}(a(s)/x(s+1)^r;\ p(s+1)) \cup \cup \mathrm{Exp}((b(s)-a(s))/x(s+1)^r;p(s+1)) \cup \cup \mathrm{Exp}((c(s)-z(s)a(s))/z(s+1)x(s+1)^{r+1};p(s+1))$$

And thus, looking at (1.3.4.2), one has that

(1.3.7.5) $$\left(\frac{h+i+j-r}{r-j}, \frac{i-l}{r-j}\right) \in \Delta(s+1)$$

or $$\left(\frac{h+i+j-r}{r-j}, \frac{i-l+1}{r-j}\right) \in \Delta(s+1)$$

But

(1.3.7.5) $$\frac{h+i+j-r}{r-j} = \delta(\Delta^*(0))-1, \quad \frac{i-l}{r-j} = \beta(s) - \frac{l}{r-j}$$
$$(i-l+1)/(r-j) = \beta(s) - (l-1)/(r-j).$$

On the other hand, the main vertex of $\Delta^{*\prime}(1)$ is given by

(1.3.7.6) $$(\alpha^{*\prime}(\Delta^*(1)) + \beta^{*\prime}(\Delta^*(1)-1),\ \beta^{*\prime}(\Delta^*(1))) =$$
$$= (\delta(\Delta^*(0)) - 1,\ \beta^{*\prime}(\Delta^*(1))).$$

So, looking at (1.3.7.4) and (1.3.7.5) and applying the lemma (1.3.4) one deduces that the ordinate $\beta(s+1)$ of the main vertex of $\Delta(s+1)$ verifies

(1.3.7.7) $$\beta(s+1) \leq \beta(s) - (l-1)/(r-j) \leq \beta(s) < \beta(0)$$

and the proof of (1.1.7) is finished.

2. NO STANDARD TRANSITIONS FROM TYPE I

(2.1) Introduction

(2.1.1) Here we shall suppose that $(X,E,\mathcal{D},P)$ is of the type I-2 or I-1-0 and that $p(0) = (x(0),y(0),z(0))$ is a very well prepared system of regular parameters. Now we shall suppose that the player A make its choice always following a l-retarded standard winning strategy, i.e.:

i) If $(x(0),z(0))$ is permissible, then A chooses this center.

ii) If $(x(0),z(0))$ is not permissible and $(y(0),z(0))$ is permissible and $\Delta(\mathcal{D}(0),E(0),p(0))$ has only one vertex, then the player A chooses the center $(y(0),z(0))$ or the quadratic center.

iii) Otherwise A chooses the quadratic center.

(2.1.2) Let $\pi(1)\colon X(1) \longrightarrow X(0)$ be the directional blowing-up which has been chosen by the player B with the center designed by A. Let $(X(1),E(1),\mathcal{D}(1),P(1))$ be the strict transform of $(X(0),E(0),\mathcal{D}(0),P(0))$. Assume that this transition is not standard. One has the following four possibilites:

a) $e(E(1)) = 1$ and $\pi(1)$ quadratic

b) $e(E(1)) = 2$ and $\pi(1)$ quadratic

c) $\pi(1)$ monoidal with center $(x(0),z(0))$

c) $\pi(1)$ monoidal with center $(y(0),z(0))$.

(2.1.3) This section is devoted to prove that the player A can win in the cases a), b) and d) above. Moreover, in the case c), the player A (if he does not win) can obtain a "bridge type".

The computations in this section will be also useful for the study of the bridge type.

(2.2) Preliminaries for the case $e(E(1)) = 1$ and $\pi(1)$ quadratic

(2.2.1) lemma. Without loss of generality, one can assume that

i) $\pi(1)$ is given by (T-1,0) from $p(0)$.

ii) $p(0)$ is well prepared (but not necessarily very well prepared).

iii) $(x(0),z(0))$ is not permissible.

Proof. Since $\underline{z}(0) \in J(\mathcal{D},E)$, if the player A has not won, $\pi(1)$ is given by $(T-1,\zeta)$ or T-2 from $p(0)$, and T-2 is not possible since $e(E(1)) = 1$. Now, one can substitue $p(0)$ by the result of making a change $y_1 = y + \zeta x$ and a good preparation: this gives i) and ii). Now iii) follows from the fact that the abscissa of the main vertex remains unchanged.

(2.2.2) First, let us remark that $(X,E,\mathcal{D},P)$ must be of the type I-2, since if one has type I-1-0, then $J(\mathcal{D},E) = (\underline{x},\underline{z})$ and the adapted order drops after T-1,0. Assume that $\mathcal{D}$ is generated by

(2.2.2.1) $$D = ax\partial/\partial x + b\partial/\partial y + c\partial/\partial z$$

where $\nu(c) \geq r+1$, $In(b) = \underline{z}^r$. Then, after T-1,0, $\mathcal{D}(1)$ is generated by

(2.2.2.2) $$D(1) = z(1)^r\partial/\partial y(1) + x(1)D^*(1)$$

$$D^*(1) = a^*(1)x(1)\partial/\partial y(1) + b^*(1)\partial/\partial y(1) + c^*(1)\partial/\partial z(1).$$

If the player A has not won, then $\nu(D^*(1),E(1)) \geq r-1$. If $\nu(c^*(1)) \geq r$ and the player A has not won, then one has a standard transition I $\longrightarrow$ I. Thus, one can assume that $\nu(c^*(1)) = r-1$.

(2.2.3) One has that $J_H(\mathcal{D}(1),E(1)) \ni \underline{x}(1)$ and then or the player A has won, or one has type zero (player A wins too) or

(2.2.3.1) $$J_H(\mathcal{D}(1),E(1)) = (\underline{x}(1))$$

and one has type 4-0 (i.e. not transversal). In the sequel let us assume that one has this case.

(2.2.4) Lemma. We can suppose that there is no permissible centers tangents to the directrix in $(X(1),E(1),\mathcal{D}(1),P(1))$ since otherwise the player A will win in the following movement.

Proof. $J(\mathcal{D}(1),E(1)) = (\underline{x}(1),\underline{z}(1))$. In view of (2.2.2.2), the only curve in Sing $(\mathcal{D}(1),E(1))$ tangent to the directrix and contained in E(1) is $x(1) = z(1) = 0$. Since by (2.2.3.1) one has

(2.2.4.1) $$D(1) = a(1)x(1)\partial/\partial x(1) + b(1)\partial/\partial y(1) + c(1)\partial/\partial z(1)$$

with $\text{In}\,(c(1)) = \lambda\underline{x}(1)^r$ $(\lambda\neq 0)$, $\text{In}\,(b(1)) = \underline{z}(1)^r + \underline{x}(1)(\ldots)$ if $x(1) = z(1) = 0$ is permissible, then there is no point over P of adapted order = r after the monoidal blowing-up.

(2.2.5) Remark. The above "vertical" curve $x(1) = z(1) = 0$ will maybe chosen by player A, although it is not permissible, as we shall see later. For globalization purposes, it is interesting to remark that this curve is permissible if it is r-fold.

(2.2.6) Since $J(\mathcal{D}(1),E(1)) = (\underline{x}(1),\underline{z}(1))$ if the player A chooses the quadratic blowing-up, then the player B must choose the transformation T-2 from p(1). Then

(2.2.6.1) $$D(2) = z(2)^r[-x(2)\partial/\partial x(2) + y(2)\partial/\partial y(2) - z(2)\partial/\partial z(2)] + x(2)D^*(2)$$

$$D^*(2) = a^*(2)x(2)\partial/\partial x(2) + b^*(2)y(2)\partial/\partial y(2) + c^*(2)\partial/\partial z(2).$$

If the player A has not won, then

$$\nu(D^*(2).E(2)) \geq r-1$$

Moreover $(1,0,0) \in \mathrm{Exp}\,(x(2)c^*(2);p(2))$.

If $(X(2),E(2),\mathcal{D}(2),P(2))$ is not of the type zero (or better) one can assume that

(2.2.6.2) $$\mathrm{In}\,(c^*(2)) = \underline{x}(2).\psi(\underline{x}(2),\underline{y}(2)).$$

Thus $(X(2),E(2),\mathcal{D}(2),P(2))$ must be of the type one, type I'-2-2 or I'-1-0. Let us denote

(2.2.6.3) $$p'(2) = (x'(2),y'(2),z'(2)) = (y(2),z(2),x(2)).$$

(2.2.7) Theorem. With notations as above, let us consider the r-1 following movements of the reduction game from the status $(X(2),E(2),\mathcal{D}(2),P(2))$. Then, if each time the player A chooses the quadratic center, necessarily the player B must choose the directional blowing-up given inductively by (T-1,0) from $p'(2)$; otherwise the player A will win.

Proof. Let $0 \leq t \leq r-1$ and let $D(t+2)$ be the result of applying t-times T-1,0 to $D(2)$. Assume that $r = \nu(\mathcal{D}(t+2),E(t+2))$. In order to simplify notation, let us write $(u,v,w) = (x'(t+2),y'(t+2),z'(t+2))$. From (2.2.6) one has

(2.2.7.1) $$D(t+2) = u^t v^r [u\partial/\partial u - (t+1)v\partial/\partial v - (t+1)w\partial/\partial w] + wD^{*\prime}(t+2)$$

$$D^{*\prime}(t+2) = a^{*\prime}(t+2)u\partial/\partial u + b^{*\prime}(t+2)\partial/\partial v + c^{*\prime}(t+2)w\partial/\partial w$$

where

(2.2.7.2) $$\nu(D^{*\prime}(t+2),E(t+2)) = r-1$$

$$(0,0,r) \in \mathrm{Exp}\,(wb^{*\prime}(t+2);p'(t+2)).$$

Then, $\underline{w} \in J(\mathcal{D}(t+2),E(t+2))$ and the next quadratic transformation is given from $p'(t+2)$ by $(T-1,0)$, $(T-1,\zeta)$, $\zeta\neq 0$, or T-2. If it is given by $(T-1,0)$ then the induction continues. Then one has to prove that the player A can win if the player B chooses $(T-1,\zeta)$, $\zeta\neq 0$, or T-2.

<u>First case</u>: <u>transformation</u> $(T-1,\zeta)$, $\zeta\neq 0$. Let $v_1 = v+\zeta u$, then

(2.2.7.3) $$D(t+2) = u^t(v_1-\zeta u)^r[u\partial/\partial u+(\zeta u-(t+1)(v_1-\zeta u))\partial/\partial v_1 - -(t+1)w\partial/\partial w]+wD^{*\prime}(t+2).$$

If one makes $(T-1,0)$ from (u,v_1,w), then

(2.2.7.4) $$D(t+3) = u'^{t+1}(v'_1-\zeta)^r[u'\partial/\partial u-(t+2)(v'_1-\zeta)\partial/\partial v'_1 - -(t+2)w'\partial/\partial w']+w'D^{*\prime}(t+3).$$

and since $t+1 \leq r-1$, the adapted order has dropped.

<u>Second case</u>: <u>T-2</u>. Here the player A wins by choosing always the quadratic center. Let $u_o,s_1,u_1,\ldots,s_l,u_l$ be nonegative integers with, $u_o = 1$, $(s_i,u_i)\neq(0,0)$, $i=1,\ldots,l$. Let $n=u_o+s_1+u_1+\ldots+u_l$ and $h=t+2+n$. Let $D(h)$ be the result of applying to $D(t+2)$, $u_o=1$ transformations T-2, s_1 transformations $(T-1,0)$, u_1 transformations T-2,...,n_l transformations T-2, inductively from $p'(t+2)$.

Let $(u,v,w) = (x'(h),y'(h),z'(h))$. If $\nu(\mathcal{D}(h),E(h)) = r$, then

(2.2.7.5) $$D(h) = u^{e(l)}v^{f(l)}[\lambda_l u\partial/\partial u+\mu_l v\partial/\partial v+\gamma_l w\partial/\partial w]+wD^{*\prime}(h)$$

Where $e(l),f(l)$ are obtained inductively by

(2.2.7.6) $$(e(0),f(0)) = (t,t+1)$$
$$(e(l),f(l)) = \sigma_2^{u_l}\sigma_1^{s_l}(e(l-1),f(l-1))$$

where $\sigma_1(x,y) = (x+y-r,y)$, $\sigma_2(x,y) = (x,x+y-r)$. The numbers λ_l,μ_l,γ_l are obtained inductively by

(2.2.7.7) $$\lambda_o = t+2,\ \mu_o = -(t+1),\ \gamma_o = 0$$
$$\lambda_l = \lambda_{l-1} - u_l\mu_l$$

$$\mu_l = \mu_{l-1} - s_l\lambda_{l-1}$$

$$\gamma_l = \gamma_{l-1} - s_l\lambda_{l-1} - u_l\mu_l$$

Let us denote

(2.2.7.8) $$D^{*\prime}(h) = a^{*\prime}(h)u\partial/\partial u + b^{*\prime}(h)v\partial/\partial v + c^{*\prime}(h)w\partial/\partial w$$

Now, since

(2.2.7.9) $$\mathrm{In}(b^{*\prime}(2)) = \underline{z}'(2)^{r-1} + \underline{x}'(2)(\ldots)$$

(see (2.2.6)), one deduces that

(2.2.7.10) $$\mathrm{In}^{r-1}(a^{*\prime}(h)) = \lambda\delta_l\underline{w}^{r-1} + \phi_1$$

$$\mathrm{In}^{r-1}(b^{*\prime}(h)) = \lambda\varepsilon_l\underline{w}^{r-1} + \phi_2$$

$$\mathrm{In}^{r-1}(c^{*\prime}(h)) = \lambda\xi_l\underline{w}^{r-1} + \phi_3$$

where $\phi_i(0,0,\underline{w}) = 0$, $i=1,2,3$ and $\delta_l, \varepsilon_l, \xi_l$ are given inductively by

(2.2.7.11) $$\delta_o = -1 \; ; \; \varepsilon_o = 1 \; ; \; \xi_o = -1$$

$$\delta_l = \delta_{l-1} - u_l\varepsilon_l$$

$$\varepsilon_l = \varepsilon_{l-1} - s_l\delta_{l-1}$$

$$\xi_l = \xi_{l-1} - s_l\delta_{l-1} - u_l\varepsilon_l.$$

One can prove that $(\lambda_l, \mu_l) \neq (0,0) \neq (\delta_l, \varepsilon_l)$ and that

(2.2.7.12) $$e(l) + f(l) \leq e(l-1) + f(l-1) - (s_l+u_l)$$

(2.2.7.13) $$e(l) \leq e(l-1); \; f(l) \leq f(l-1)$$

(2.2.7.14) $$1 \leq e(l) \leq r-1; \; 1 \leq f(l) \leq r-1.$$

Note, for (2.2.7.14), that $e(l) = 0$ implies $e(l) + f(l) < r$ (may be proved by induction) in contradiction with the fact $\nu(\mathcal{D}(h), E(h)) = r$.

As a corollary of (2.2.7.12) one has that

(2.2.7.15) $$r \leq e(l)+f(l) \leq 2t+1-n$$

and so $n \leq 2t+1$.

Now, let us prove that if the player A chooses the quadratic center in $(X(h),E(h),\mathcal{D}(h),P(h))$, then the player B must choose $(T-1,0)$, $T-2$ or $(T-1,\zeta)$, $\zeta\neq 0$ and if he chooses $(T-1,\zeta)$, $\zeta\neq 0$, then the player A wins. Since the player B cannot choose always $(T-1,0)$ or $T-2$ in view of $n\leq 2t+1$, then the player A will always win by choosing the quadratic center.

First, let us prove that one can suppose that

$$e(1)+f(1)\geq r+1. \tag{2.2.7.16}$$

If $e(1)+f(1)=r$, one has two possibilities (unless symmetry): 1º. $\delta_1\neq 0\neq\lambda_1$, 2º. $\delta_1\neq 0\neq\mu_1$, $\lambda_1=0$. In the first case

$$\begin{aligned} In^r(\lambda_1 u^{e(1)}v^{f(1)}+wa^{*\prime}(h)) &= \lambda\delta_1\underline{w}^r+ \\ &+\underline{w}\phi_1(\underline{u},\underline{v},\underline{w})+\lambda_1\underline{u}^{e(1)}\underline{v}^{f(1)} \end{aligned} \tag{2.2.7.17}$$

and since $e(1)\geq 1$, $f(1)\geq 1$ the directrix of this coefficient is of dimension zero. In the second case,

$$\begin{aligned} &\underline{w}\in J^r(wa^{*\prime}(h)) \\ &In^r(\mu_1 u^{e(1)}v^{f(1)}+wb^{*\prime}(h))=\mu_1\underline{u}^{e(1)}\underline{v}^{f(1)} \qquad (\mathrm{mod}\ \underline{w}) \end{aligned} \tag{2.2.7.18}$$

and since $e(1)\geq 1$, $f(1)\geq 1$, $\dim \mathrm{Dir}\,(\mathcal{D}(h),E(h))=0$. Then (2.2.7.16) may be supposed true.

Now, by (2.2.7.16) and looking at (2.2.7.5) one deduces that $\underline{w}\in J(\mathcal{D}(h),E(h))$ and the player B must choose $(T-1,0)$, $T-2$ or $(T-1,\zeta)$, $\zeta\neq 0$. Now, applying $T-1,0$ to $(u,v_1=v+\zeta u,w)$ one has

$$\begin{aligned} D(h+1) &= u'^{\,e(1)+f(1)-r}(v'_1-\zeta)^{f(1)}(\lambda_1 u'\partial/\partial u'+ \\ &+(\mu_1-\lambda_1)(v'_1-\zeta)\partial/\partial v'_1+(\gamma_1-\lambda_1)w'\partial/\partial w')+w'D^{*\prime}(h+1) \end{aligned} \tag{2.2.7.19}$$

and since $e(1)+f(1)<2r-1$ by (2.2.7.14) the order has dropped.

(2.3) The case $e(E(1))=1$ and $\pi(1)$ quadratic

(2.3.1) Let us assume along this paragraph that $(X(2),E(2),\mathcal{D}(2),P(2))$ is obtained from $(X,E,\mathcal{D},P)$ as in (2.2), that the player A has not won in this transition and that if the player A chooses r−1 times the quadratic center, then he does not win.

More precisely, let us denote by

(2.3.1.1) $$(X(r+1),E(r+1),\mathcal{D}(r+1),P(r+1))$$

the result of applying r−1 times the transformation (T−1,0) from p'(2) to $(X(2),E(2),\mathcal{D}(2),P(2))$. Since the player A has not won, $r = \nu(\mathcal{D}(r+1,E(r+1))$, dim Dir $(\mathcal{D}(r+1),E(r+1) \geq 1$ and one has not type zero (then one has type one I').

The main result is that with the above hypothesis the player A can win by making a monoidal transformation centered at $(x(1),z(1))$ in $(X(1),E(1),\mathcal{D}(1),P(1))$ and thus he can avoid the above situation.

(2.3.2) <u>Definition</u>. Let $p = (x,y,z)$ be a regular system of parameters suited for (E,P) in a situation $(X,E,\mathcal{D},P)$ such that Exp (D,E;p) is defined. Given a subset $A \subset \{1,2,3\}$, the "order $\underline{\nu}_A(\mathcal{D},E,p)$" is

(2.3.2.1) $$\underline{\nu}_A(\mathcal{D},E,p) = \inf \{ \sum_{i \in A} h_i ; (h_1,h_2,h_3) \in \text{Exp } (\mathcal{D},E,p)\}.$$

(2.3.3) <u>Remark</u>. If vgr. Y is given by (y,z) and A = (2,3), in general $\underline{\nu}_A(\mathcal{D},E,p)$ is not the generic order along Y. In this case $\underline{\nu}_A(\mathcal{D},E,p)$ may be considered as an invariant which says "how far is Y from being permissible". More precisely, Y is permissible iff

(2.3.3.1) $$\underline{\nu}_A(\mathcal{D},E,p) = r.$$

(2.3.4) Let us recall that in view of (2.2.6), one has type one I'-2-2 or I'-1-0 for $(X(2),E(2),\mathcal{D}(2),P(2))$, then if $\mathcal{D}(2)$ is generated by

(2.3.4.1) $$D(2) = a'(2)x'(2)\partial/\partial x'(2) + b'(2)\partial/\partial y'(2) + c'(2)z'(2)\partial/\partial z'(2)$$

Let us consider

(2.3.4.2) $$\text{Exp } (D(2),E(2),p'(2)) = \text{Exp } (y'(2)a'(2);p'(2)) \cup \cup \text{Exp } (b'(2);p'(2)) \cup \text{Exp } (y'(2)c'(2);p'(2)).$$

The Exp is defined as for $\mathcal{D}(2)$ relatively to $x(2) = 0$ if it is of the type I. Analogously, the polygon $\Delta(\mathcal{D}(2),E(2);p(2))$ may be defined by the projection from $(0,0,r)$ in the usual way.

(2.3.5) Lemma. There is a bijection

(2.3.5.1) $$\phi\colon \mathrm{Exp}\,(D(0),E(0);p(0)) \longrightarrow \mathrm{Exp}\,(D(2),E(2);p(2))$$

given by

(2.3.5.2) $$\phi(h,i,j) = (h+2(i+j-r),j+1,h+i+j-r)$$

$$\phi^{-1}(h',i',j') = (2j'-h',r+h'-i'-j'+1,i'-1).$$

Proof. It follows from the following equations:

(2.3.5.3)
$$\begin{array}{lll} x(1) = x(0) & x(2) = x(1)y(1) & x'(2) = y(2) \\ y(1) = y(0)x(0) & y(2) = y(1) & y'(2) = z(2) \\ z(1) = z(0)x(0) & z(2) = z(1)y(1) & z'(2) = x(2) \end{array}$$

and

(2.3.5.4)
$$a(1) = a(0)/x(0)^{r-1}$$

$$b(1) = b(1)/x(0)^{r}-y(1)a(0)/x(0)^{r-1}$$

$$c(1) = c(1)/x(0)^{r}-z(1)a(0)/x(0)^{r-1}$$

$$a(2) = a(1)/y(1)^{r-1}-b(1)/y(1)^{r}$$

$$b(2) = b(1)/y(1)^{r}$$

$$c(2) = c(1)/y(1)^{r}-z(2)b(1)/y(1)^{r}$$

$$a'(2) = b(2);\ b'(2) = c(2);\ c'(2) = a(2).$$

(2.3.6) Theorem. Let $A = \{2,3\}$. Assume that

(2.3.6.1) $$\underline{\nu}_A(\mathcal{D}(2),E(2);p'(2)) \leq 1.$$

Then, there is a vertex

(2.3.6.2) $$w = (u,v) \in \Delta(\mathcal{D}(2),E(2);p'(2))$$

such that

(2.3.6.3) $$v + rv \leq r \text{ and } v < 1.$$

Moreover, if $r \geq 3$, one can take w such that

(2.3.6.4) $$u + rv < r.$$

Proof. It must to exist

(2.3.6.5) $$(h',i',j') \in \mathrm{Exp}\ (\mathcal{D}(2),E(2);p'(2))$$

such that $i'+j' \leq 1$. One has that $j' \leq 1$, and then there exists a point

(2.3.6.6) $$w' = (h'/(r-j');\ i'/(r-j')) \in \Delta(\mathcal{D}(2),E(2);p'(2)).$$

Obviously, it is enough to prove (2.3.6.3) and (2.3.6.4) for this point w'. Let

(2.3.6.7) $$(h,i,j) = \phi^{-1}(h',i',j') \in \mathrm{Exp}\ (\mathcal{D}(0),E(0);p(0))$$

(by lemma (2.3.5)). One has that

(2.3.6.8) $$i'+j' \leq 1 \Leftrightarrow j+1+h+i+j-r \leq 1 \Leftrightarrow h' \leq i-r.$$

(see Lemma (2.3.5)). Now, let us prove that $i - r \leq r$. If $i-r \geq r+1$, one has that

(2.3.6.9) $$i \geq 2r+1 \Rightarrow j' = h+i+j-r \geq 2r+1+h+j-r \Rightarrow j' \geq r+1+h+j \geq r$$

(Remark that $0 \leq r$, $-1 \leq j$) but this contradicts $i'+j' \leq 1$. So, $h' \leq r$

Let $r \geq 3$. One has that

(2.3.6.10) $$h'/(r-j') + ri'/(r-j') < r$$

iff

(2.3.6.11) $$h' < r(r-(i'+j')).$$

but this is always true since $r \geq 3$ and $i'+j' \leq 1$. The same argument proves the first part of (2.3.6.3) for $r = 2$. Finally, let us suppose that $r = 2$ and

(2.3.6.12) $$(h'/(r-j'),i'/(r-j')) = (0,1)$$

Then $i' = j' = 1$ and $h' = 0$. But this implies that

(2.3.6.13) $$(0,1,1) \in \mathrm{Exp}\ (b'(2);p'(2))$$

and then one has type zero for $(X(2),E(2),\mathcal{D}(2),P(2))$, contradiction.

(2.3.7) Proposition. Assume that

(2.3.7.1) $$\underline{\nu}_A(\mathcal{D}(2),E(2);p'(2)) \geq 2.$$

Then if the player A chooses the center (x(1),z(1)) for (X(1),E(1),$\mathcal{D}$(1),P(1)) he wins in the sense that the adapted order will drop for each point over P(1).

Proof. Let $r' = \underline{\nu}_A(\mathcal{D}(2),E(2);p'(2))$ and let $Y' \subset X(2)$ be given by (y'(2),z'(2)) and $Y \subset X(1)$ be given by (x(1),z(1)). One has that

(2.3.7.2)
$$\nu_{Y'}(a'(2)) \geq r'-1$$
$$\nu_{Y'}(b'(2)) \geq r'$$
$$\nu_{Y'}(c'(2)) \geq r'-1$$

looking at the equations in the proof of (2.3.5) one deduces that

(2.3.7.3) $$\nu_Y(a(1)) \geq r'-1;\ \nu_Y(b(1)) \geq r'-1;\ \nu_Y(c(1)) \geq r',$$

and moreover r' is the maximum number which satisfies the above inequalities. First let us suppose that the player B chooses the transformation

(2.3.7.4)
$$x(1) = x(2)$$
$$y(1) = y(2)$$
$$z(1) = (z(2)+\zeta)x(2)$$

then the strict transformation is given by

(2.3.7.5)
$$D(2) = (1/x(2))^{r'-1}(a(1)x(2)\partial/\partial x(2)+b(1)\partial/\partial y(2) +$$
$$+ (c(1)/x(2) - (z(2)-\zeta)a(1))\partial/\partial z(2)).$$

And the adapted order drops since

(2.3.7.6) $$\nu(c(1)/x(2)^{r'}) \leq r-r' < r$$

because

(2.3.7.7) $$\mathrm{In}^r(c(1)) = \underline{x}(1)^r$$

(see (2.2.3.1)).

If the player B chooses the transformation

$$(2.3.7.8) \qquad x(1) = x(2)z(2);\; y(1) = y(2);\; z(1) = z(2)$$

then the strict transform is given by

$$(2.3.7.9) \qquad D(2) = (1/z(2))^{r'-1}((a(1)-c(1)/z(2))x(2)\partial/\partial x(2) + $$
$$+ b(1)\partial/\partial y(2) + (c(2)/z(2))z(2)\partial/\partial z(2)).$$

And the adapted order drops since

$$(2.3.7.10) \qquad \nu(b(1)/z(2)^{r'-1}) \leq r-r'+1 < r$$

because

$$(2.3.7.11) \qquad b(1) = z(1)^r + x(1)(\ldots)$$

(see (2.2.2.2)).

(2.3.8) Proposition. With notations as above, if

$$(2.3.8.1) \qquad \underline{\nu}_{\{2,3\}}(X(2),E(2),\mathcal{D}(2),P(2)) \leq 1$$

then

a) If $r \geq 3$ then the player A wins by choosing the quadratic center at most in the following $r-1$ movements from $(X(2),E(2),\mathcal{D}(2),P(2))$.

b) If $r=2$, then the player A wins by choosing the quadratic center at most in the following $r+1$ movements from $(X(2),E(2),\mathcal{D}(2),P(2))$.

Proof. In view of the theorem (2.2.7), one can suppose that B has chosen $r-1$ times the directional blowing-up given inductively by $(T-1,0)$ from $p'(2)$. Let us denote by

$$(2.3.8.2) \qquad \Delta(i) = \Delta(\mathcal{D}(i),E(i);p'(i))$$

$i=2,3,\ldots,r+1$. It is clear that

$$(2.3.8.3) \qquad \Delta(i+1) = \sigma(\Delta(i)) \qquad i=2,\ldots,r$$

where $\sigma(x,y)=(x+y-1,y)$ (always under the hypothesis that $\nu(\mathcal{D}(i),E(i),P(i)) = r$, $2 \leq i \leq r$).

Now, in view of the theorem (2.3.6) one has that if $r \geq 3$:

(2.3.8.14) $$\Delta(r+1) \cap \{(u,v);\ u+v < 1\} \neq \emptyset$$

and this implies that $\nu(\mathcal{D}(r+1),E(r+1),P(r+1)) < r$. This proves part a).

Let us suppose that $r = 2$. First, let us observe that in view of (2.2.6.1) the only vertex of $\Delta(2)$ with no entire coordinates is (0,3/2). Moreover, the vertex (0,1) does not appear since otherwise one has the type zero. Then, if (2.3.8.4) is not true, by the theorem (2.3.6) one has that $(0,2) \in \Delta(2)$. This implies that $(0,1) \in \Delta(3)$. But since then the initial form of b(3) is divisible by z'(3) (make the computations from (2.2.6.1)), the directrix of $(X(3),E(3),\mathcal{D}(3),P(3))$ is given by $J(\mathcal{D}(3),E(3)) = (\underline{x}'(3),\underline{z}'(3))$.

So the player B must choose T-2, after making T-2, one has that (½,1) and (0,1) are the only vertices of $\Delta(4)$ and easy computations from (2.2.6.1) show that the directrix of $(X(4),E(4),\mathcal{D}(4),P(4))$ is given by $(\underline{x}'(4),\underline{z}'(4))$. So the player B must choose T-2 once more. Then (½,½) and (0,1) are the only vertices of $\Delta(5)$. If follows easily that

(2.3.8.7) $$\mathrm{In}\ (b'(5)) = \lambda\underline{y}'(5)\underline{x}'(5) + \mu\underline{z}'(5)^2 + \gamma\underline{z}'(5)\underline{x}'(5)$$

where $\lambda \neq 0, \gamma\ \mu \neq 0$. So dim Dir $(\mathcal{D}(5),E(5)) = 0$ and the player A has won.

(2.4) The case e(E(1)) = 2 and $\pi(1)$ quadratic

(2.4.1) In this section, we shall treat the possibility b) of the introduction (2,1). As we have done in the case of e(E(1)) = 1, we shall give a direct way for the player A to win by using weakly permissible curves which are not permissible. In view of the results of (III. (3.2) and (3.3)) and the remark (III.(3.3.7)), it is enough to consider here the case that $(X,E,\mathcal{D},P)$ is of the type I-2.

(2.4.2) In the situation of (2.1), $\pi(1)$ is given by (T-2) from p(0). Assume that $\mathcal{D}(0)$ is generated by

(2.4.2.1) $$D = a(0)x(0)\partial/\partial x(0) + b(0)\partial/\partial y(0) + c(0)\partial/\partial z(0),$$

then, $\mathcal{D}(1)$ is generated by

$$(2.4.2.2)\qquad D(1) = z(1)^r(-x(1)\partial/\partial x(1) + y(1)\partial/\partial y(1) - z(1)\partial/\partial z(1)) + y(1)D^*(1) = a(1)x(1)\partial/\partial x(1) + b(1)y(1)\partial/\partial y(1) + c(1)\partial/\partial z(1).$$

(recall that $In(b(0)) = \underline{z}(0)^r$). Since the transition is not standard, one has that

$$(2.4.2.3)\qquad \nu(c(1)) = r.$$

Since $\underline{y}(1) \mid In\ (c(1))$ one has the following possibilities for the directrix:

$$(2.4.2.4)\qquad \begin{aligned} J(\mathcal{D}(1),E(1)) &= (\underline{y}(1)) \\ \ldots\quad \ldots &= (\underline{y}(1),\underline{x}(1)) \\ \ldots\quad \ldots &= (\underline{y}(1),\underline{z}(1) + \lambda\underline{x}(1)) \\ \ldots\quad \ldots &= (\underline{y}(1),\underline{z}(1),\underline{x}(1) \end{aligned}$$

The last case has dim Dir $(\mathcal{D}(1),E(1)) = 0$ and the third case is of the type zero. In the second case, the player A chooses the quadratic center, then the player B must choose the transformation

$$(2.4.2.5)\qquad y(2) = y(1)z(1);\ x(2) = x(1)z(1);\ z(2) = z(1)$$

and (see (2.4.2.2)) one has that

$$(2.4.2.6)\qquad \nu(\mathcal{D}(2),E(2),P(2)) < r$$

Thus, let us assume that

$$(2.4.2.7)\qquad J(\mathcal{D}(1),E(1)) = (\underline{y}(1)).$$

(2.4.3) Lemma. Let $p'(1) = (x'(1),y'(1),z'(1)) = (x(1),z(1),y(1))$. Then $(X(1),E(1),\mathcal{D}(1),P(1))$ is of the type I'-2-2 and $p'(1)$ is normalized. Moreover, there is a bijection

$$(2.4.3.1)\qquad \phi : Exp(D(0),E(0),p(0)) \longrightarrow Exp(D(1),E(1),p'(1))$$

given by

(2.4.3.2) $$\phi(h,i,j) = (h,j+1,h+i+j-r)$$

$$\phi^{-1}(h',i',j') = (h',j'-h'-i'+r+1,i'-1)$$

Proof. The first part is trivial. For the second part it is enough to look at the equations:

(2.4.3.3) $$D(1) = a'(1)x'(1)\partial/\partial x'(1) + b'(1)\partial/\partial y'(1) + c'(1)z'(1)\partial/\partial z'(1)$$

(2.4.3.4) $$x(1) = x(0)y(0);\ y(1) = y(0);\ z(1) = z(0)y(0)$$

(2.4.3.5) $$a(1) = a(0)/y(0)^{r-1} - b(0)/y(0)^r$$

$$b(1) = b(0)/y(0)^r$$

$$c(1) = c(0)/y(0)^r - z(1)b(0)/y(0)^r$$

(2.4.3.6) $$a'(1) = a(1);\ b'(1) = c(1);\ c'(1) = b(1).$$

(Recall that

(2.4.3.7) $$\mathrm{Exp}\ (D(1),E(1),p'(1)) = \mathrm{Exp}\ (a'(1)y'(1);p'(1))\cup$$
$$\cup \mathrm{Exp}\ (b'(1);p'(1)) \cup \mathrm{Exp}\ (c'(1)y'(1);p'(1))$$

since one has type I'-2).

(2.4.4) Theorem. In the above situation, there are two possibilities:

a) The player A wins by choosing the curve $(y'(1),z'(1))$ as center.

b) The player A wins by choosing a permissible center in less than three steps.

Proof. The proposition (2.3.7) can be applied to $(X(1),E(1),\mathcal{D}(1),P(1))$ and $p'(1)$. So if a) is not true we can suppose that

(2.4.4.1) $$\underline{\nu}_{\{2,3\}}(\mathcal{D}(1),E(1),p'(1)) \leq 1.$$

Then, there exist $(h',i',j') \in \mathrm{Exp}\ (D(1),E(1),p'(1))$ such that

(2.4.4.2) $$i' + j' \leq 1.$$

First, let us observe that in view of (2.4.2.2), one has the same property that in the theorem (2.2.7) and one can assure that if the player A chooses the quadratic center, then the player B must choose (T-1,0) at least r-1 times from p'(1).

Let $(h,i,j) = \phi^{-1}(h',i',j')$. By (2.4.3.7), one has that $0 \leq i',j'$. Then there are two possibilities

(2.4.4.3) $$j = i'-1 = 0 \quad \text{or} \quad j = i'-1 = -1.$$

Assume j = 0, then i' = 1 and j' = 0. Then 0 = j' = h+i+j-r and

(2.4.4.4) $$h = r-i.$$

Let us observe that

(2.4.4.5) $$h' + i' + j' \geq r$$
$$h' + i' + j' = r \Rightarrow (h',i',j') = (0,0,r)$$

since one has type I'-2. So $h' + j' + i' \geq r+1$ and $h' \geq r$. Necessarily h' = h = r, i = 0. this implies that

(2.4.4.6) $$(h',i',j') = (r,1,0)$$
$$(1,1/r) \in \Delta(\mathcal{D}(1),E(1);p'(1)).$$

Now, if $r \geq 3$, the order drops in the next blowing-up (which may be supposed T-1,0) If r = 2, after the first transformation T-1,0, one has that

(2.4.4.7) $$(½,½) \in \Delta(\mathcal{D}(2),E(2);p'(2))$$

(if the adapted order has not dropped), but this implies easily that we have type zero or dim Dir $(\mathcal{D}(2),E(2)) = 0$, since $J(\mathcal{D}(2),E(2)) \not\subset (\underline{x}'(2),\underline{z}'(2))$.

Now, assume j = -1. Then i' = 0. If j' = 0, we have 0 = j' = h+i+j-r, then h = r+1-1. But since j = -1, one has that $i \geqq 1$, thus $h \leq r$, this implies

(2.4.4.8) $$(h',i',j') = (h,0,0)$$

which contradicts (2.4.4.5). Then, necessarily j' = 1. As above, one has that h' = = $h \leq r$ and 1 = h+i+j-r, so h' = r+2-i and since $1 \leq i$ one has that h'= r+1 or h' =

= r. Then

$$(1 + \frac{2}{r-1}, 0) \in \Delta(\mathcal{D}(1), E(1); p'(1)) \quad (2.4.4.9)$$

or

$$(1 + \frac{1}{r-1}, 0) \in \Delta(\mathcal{D}(1), E(1); p'(1)) \quad (2.4.4.10)$$

If $r \geq 3$, in both cases the adapted order drops in the two first transformations (T-1,0).

Assume that $r = 2$ and that $(h',i',j') = (r,0,1)$ or $(h',i',j') = (r+1,0,1)$ If $(h',i',j') = (r,0,1)$, then, after making (T-1,0), one has that

$$(1,0) \in \Delta(\mathcal{D}(2), E(2); p'(2)) \quad (2.4.4.11)$$

Moreover, the initial form of $b'(2)$ is divisible by $\underline{z}'(2)$, so $J(\mathcal{D}(2),E(2)) = (\underline{z}'(2), \underline{x}'(2))$. The other vertex of $\Delta(\mathcal{D}(2),E(2);p'(2))$ is $(1/2, 3/2)$ (the only possible point with no entire coordinates). The next two transformations must be given by T-2 and

$$(1/2, 1/2) \in \Delta(\mathcal{D}(4), E(4); p'(4)) \quad (2.4.4.12)$$

which implies type zero (or better in fact dim Dir $(\mathcal{D}(4),E(4)) = 0$).

If $(h',i',j') = (r+1,0,1)$ and $(r,0,1) \notin \text{Exp}\,(D(1),E(1);p'(1))$, the only point of no entire coordinates of the polygon is $(0,3/2)$, then one has that the only vertices of $\Delta(\mathcal{D}(1),E(1);p'(1))$ are $(0,3/2)$ and $(3,0)$. After making (T-1,0), $J(\mathcal{D}(2),E(2)) = (\underline{z}(2))$ and the vertices of $\Delta(\mathcal{D}(2),E(2);p'(2))$ are $(1/2,3/2)$ and $(2,0)$. If the next transformation is $(T-1,\zeta)$, then $(x'(3),z'(3))$ becomes permissible and after the blowing-up with this center, the adapted order drops, if the next transformation is T-2, then $(y'(3),z'(3))$ becomes permissible and after the blowing-up with this center, the adapted order drops.

(2.5) <u>$\pi(1)$ monoidal with center $(x(0),z(0))$</u>.

(2.5.1) First, let us introduce the definition of the "bridge type". In this paragraph the main result assures that if the player A does not win, then he can obtain a bridge type.

Let us assume that the initial situation of (2.1) is of the type I-2, since if one has type I-1-0, then the player A wins after the blowing-up with the center (x,z).

(2.5.2) Definition. We shall say that $(X,E,\mathcal{D},P)$ is of the "bridge type" iff

a) It is of the type one I'-2 or I'-1-0. (See chapter III, (1.2.5) and (3.5.3)).

b) There is a normalized base $p = (x,y,z)$ (see chapter III, (3.5.9)) such that $(0,1+1/r)$ is the main vertex of $\Delta(\mathcal{D},E;p)$.

(2.5.3) Remark. In terms of coordinates, conditions a) and b) above are equivalent to the existence of p such that $E = (xz=0)$ and if $\mathcal{D}$ is generated by

(2.5.3.1) $$D = ax\partial/\partial x + b\partial/\partial y + cz\partial/\partial z$$

then $J^r(b) = (\underline{z} + \lambda\underline{x})$ or $J^r(b) = (\underline{z},\underline{x})$ and

(2.5.3.2) $$(0,0,r) \in \mathrm{Exp}\,(b;p)$$

(2.5.3.3) $$(r+1,0,0) \in \mathrm{Exp}\,(ya;p) \cup \mathrm{Exp}\,(b;p) \cup \mathrm{Exp}\,(yc;p).$$

(2.5.4) lemma. Assume that the player A has not won in the status $(X(1),E(1),\mathcal{D}(1),P(1))$, then

a) $(X(1),E(1),\mathcal{D}(1),P(1))$ is of the type 4-0.

b) p(0) may be choosen in such a way that if p(1) is obtained from p(0) by T-3, then $\mathcal{D}(1)$ is generated by

(2.5.4.1) $$D(1) = a(1)x(1)\partial/\partial x(1) + b(1)\partial/\partial y(1) + c(1)\partial/\partial z(1)$$

with

(2.5.4.2) $$J^r(b(1)) = (\underline{z}(1)) \quad \text{or} \quad (\underline{z}(1),\underline{x}(1))$$
$$J^r(c(1)) = (\underline{x}(1))$$

$(0,0,r) \in \mathrm{Exp}\,(b(1);p(1))$.

Proof. Assume that $\mathcal{D}(0)$ is generated by

(2.5.4.3) $$D(0) = a(0)x(0)\partial/\partial x(0) + b(0)\partial/\partial y(0) + c(0)\partial/\partial z(0).$$

If A has not won, necessarily $\pi(1)$ is given by T-3 from p(0). Moreover, since the transition is not standard, $\nu(c(1)) = r$. Since $J^r(b(0)) = (\underline{z}(0))$ and $b(1)=b(0)/y(1)^r$ then

(2.5.4.4) $$J^r(b(1)) = (\underline{z}(1)+\alpha\underline{x}(1)+\beta\underline{y}(1))$$

or $$J^r(b(1)) = (\underline{z}(1)+\alpha\underline{x}(1)+\beta\underline{y}(1), \gamma\underline{x}(1)+\delta\underline{y}(1))$$

In both cases, making a change $z(0) \mapsto z(0) + \alpha x(0)y(0)+\beta y(0)^2$ which does not modify the hypothesis on p(0) nor the polygon $\Delta(\mathcal{D}(0),E(0);p(0))$ one can suppose that

(2.5.4.5) $$J^r(b(1)) = (\underline{z}(1)) \text{ or } (\underline{z}(1), \alpha\underline{x}(1)+\beta\underline{y}(1))$$

(Recall that if dim Dir $(\mathcal{D}(1),E(1)) = 0$ then A wins). Moreover $(0,0,r) \in$ $\in$ Exp $(b(1);p(1))$. In (2.5.4.5) above, if $\beta \neq 0$, then type zero, so $\beta = 0$.

In this situation, there are the following possibilities for $J^r(c(1))$:

(2.5.4.6) $$J^r(c(1)) = (\underline{z}(1) + \alpha\underline{x}(1) + \beta\underline{y}(1))$$
$$J^r(c(1)) = (\alpha\underline{x}(1) + \beta\underline{y}(1))$$
$$J^r(c(1)) = (\underline{z}(1), \alpha\underline{x}(1) + \beta\underline{y}(1))$$

In each case, if $\beta \neq 0$, then type zero (or better). Moreover, the first and third cases are of the type zero too. So necessarily

(2.5.4.7) $$J^r(c(1)) = (\underline{x}(1))$$

this proves b) and a) is automatic from b).

(2.5.5) Theorem. With the hypothesis of (2.1). If in the status $(X(1),E(1),\mathcal{D}(1),P(1))$ the player A has not won and he does not win in the next movement, then by choosing the quadratic center, necessarily the status $(X(2),E(2),\mathcal{D}(2),P(2))$ is of the bridge type.

Proof. The above lemma implies that

(2.5.5.1) $$J(\mathcal{D}(1),E(1)) = (\underline{x}(1),z(1))$$

So if A chooses the quadratic center, the player B must choose T-2 from p(1). Let us suppose that

(2.5.5.2) $$In\ (b(1)) = \phi(\underline{x}(1),\underline{z}(1))$$

with $\phi(0,z) = z^r$. Then, after making T-2, $\mathcal{D}(2)$ is generated by

(2.5.5.3) $$D(2) = \phi(x(2),z(2))[-x(2)\partial/\partial x(2) + y(2)\partial/\partial y(2) - - z(2)\partial/\partial z(2)] + x(2)^r\partial/\partial z(2) + y(2).D^*(2)$$

and E(2) is given by x(2).y(2). Let p'(2) = (x'(2),y'(2),z'(2)) = (y(2),z(2),x(2)). Now if $r = \nu(\mathcal{D}(2),E(2),P(2))$ and $1 \leq \dim \mathrm{Dir}\ (\mathcal{D}(2),E(2))$ then $(X(2),E(2),\mathcal{D}(2),P(2))$ and p'(2) satisfies the conditions a) and b) of (2.5.2).

2.6. $\pi(1)$ monoidal with center (y(0),z(0))

(2.6.1) In this case the player A always wins. If he does not win in the first movement he will win in the second one by choosing a quadratic center or a permissible monoidal one. For the proof of this it is important the fact that $\Delta(\mathcal{D}(0),E(0),p(0))$ has only one vertex. Also this case has sense only for the initial situation of the type I-2.

(2.6.2) If the player A has not won the player B must choose the transformation T-4 from p(0). Then $\mathcal{D}(1)$ is generated by

(2.6.2.1) $$D(1) = a(1)x(1)\partial/\partial x(1) + b(1)y(1)\partial/\partial y(1) + c(1)\partial/\partial z(1)$$

where $a(1) = a(0)/y(1)^{r-1}$, $b(1) = b(0)/y(1)^r$, $c(1) = c(0)/y(1)^r - z(1)b(0)/y(1)^r$. Since the transition is not standard, necessarily $\nu(c(1)) = r$. If one has not $\dim \mathrm{Dir}\ (\mathcal{D}(1),E(1)) = 0$ or type zero, there are the following possibilities for $J^r(c(1)) = J(\mathcal{D}(1),E(1))$:

(2.6.2.2)
$$J^r(c(1)) = (\underline{y}(1) + \alpha \underline{x}(1))$$
$$J^r(c(1)) = (\underline{x}(1))$$
$$J^r(c(1)) = (\underline{x}(1), \underline{y}(1))$$

In the third case, if A chooses the quadratic center, he wins always because of the fact that

(2.6.2.3) $$(0,0,r) \in \mathrm{Exp}\ (b(1);p(1))$$

(the adapted order drops).

(2.6.3) Let us suppose that $J^r(c(1)) = (\underline{y}(1)+\alpha\underline{x}(1))$. This implies easily that

(2.6.3.1) $$(0,2r+1,-1) \in \mathrm{Exp}\ (y(0)c(0)/z(0);p(0))$$

and then $(0,2-1/(r+1)) \in \Delta(\mathcal{D}(0),E(0);p(0))$. Moreover, this one must be the main vertex of $\Delta(0) = \Delta(\mathcal{D}(0),E(0);p(0))$ since if

(2.6.3.2) $$(0,i,j) \in \mathrm{Exp}\ (D(0),E(0);p(0)) = \mathrm{Exp}\ (0)$$

one deduces that $i+2j \geq 2r$ or $(0,i,j) = (0,2r+i,-1)$ (otherwise the adapted order will drop or $J^r(c(1)) \neq (\underline{y}(1)+\alpha\underline{x}(1))$. Thus

(2.6.3.3) $$\beta(0) = 2-1/(r+1)$$

Let us remark that $\beta(0) > 1+1/r$.

Now, in view of the general hypothesis of (2.1) the polygon $\Delta(0)$ have only one vertex. This implies that for every $(h,i,j) \in \mathrm{Exp}\ (D(0),E(0);p(0))$ one has that

(2.6.3.4) $$i \geq (2 - \frac{1}{r+1})\ (r-j)$$

In particular, if $j\neq-1$, one has $i+2j \geq 2r$ and if $j=-1$ one has $i \geq 2r+1$. Let

(2.6.3.5)
$$\mathrm{Exp}\ (1) = \mathrm{Exp}\ (a(1);p(1)) \cup \mathrm{Exp}\ (b(1);p(1)) \cup$$
$$\cup\, \mathrm{Exp}\ (c(1)/z(1);p(1))$$

It is easy to show that there is a bijection

(2.3.6.3) ϕ: Exp (0) $\longrightarrow$ Exp (1)

given by ϕ(h,i,j) = (h,i+j-r,j). By (2.6.3.4) and (2.6.3.6) one deduces that the curve given by (y(1),z(1)) is contained in $\mathrm{Sing}^r(\mathcal{D}(1),E(1))$ and thus is permissible. Now, if A chooses this center he wins since

$$(0,0,r) \in \mathrm{Exp}\ (b(1);p(1)) \qquad (2.6.3.7)$$

$$J^r(c(1)) = (\underline{y}(1))$$

(remark that necessarily $\alpha = 0$).

(2.6.4) Assume that $J^r(c(1)) = (\underline{x}(1))$. This implies that

$$(r,r+1,-1) \in \mathrm{Exp}\ (c(0).y(0)/z(0);p(0)) \qquad (2.6.4.1)$$

and since if (h,i,j) $\in$ Exp (0) one deduces easily that $h+i+2j-r \geq r$ or (h,i,j) = = (r,r+1,-1) (remark that in c(1) the only monomial of order r is $x(1)^r$), it follows that (1-1/(r+1),1) is the only vertex of $\Delta(0)$. Then, for each (h,i,j) $\in$ Exp(0) one has that

$$h \geq (1 - \frac{1}{r+1})\ (r-j). \qquad (2.6.4.2)$$

In particular, if $j \neq -1$, $h+j \geq r$ and if $j=-1$ then $h \geq r$. As in (2.6.3) it follows that (x(1),z(1)) is permissible and if the player A chooses this center then the adapted order drops.

3. NO STANDARD TRANSITIONS FROM TYPE II

(3.1) Introduction

(3.1.1) In this section we shall consider no standard transitions produced from type I but after a few standard ones. More precisely, let us fix $(X(0),E(0),\mathcal{D}(0),P(0))$ of the type I-2 or I-1-0 and p(0) a very well prepared system of regular parameters. Let us fix a realization of the reduction game of length bigger than s+1:

(3.1.1.1) $G = \{ G(t) = (\text{mov } (t), \text{stat } (t)) \}_{t=0,1,\ldots}$

such that the player A has followed in G the 1-retarded standard winning strategy with respect to p(0) until the step s (see (1.1.4)). Moreover, assume that

(3.1.1.2) $\pi(s+1): X(s+1) \longrightarrow X(s)$

defines a no standard nor natural transition and that stat (t) is of the type II for $t = 1,2,\ldots,s$.

(3.1.2) Without loss of generality, one can assume that there is a sequence p(t) , $t = 0,1,\ldots,s+1$ of regular systems of parameters for stat (t), $t = 0,1,\ldots,s+1$ in such a way that p(t) is obtained inductively from p(t-1) by (T-1,0),T-2,T-3 or T-4, and for t=s+1, may be $(T-1,\zeta)$, $\zeta \neq 0$ and moreover, each p(t) is strongly well prepared (see (1.1.6),vgr.). Thus, one can distinguish the five followings possibilities:

a) $\pi(s+1)$ quadratic given by (T-1,0), from p(s).

b) $\pi(s+1)$ " " " $(T-1,\zeta)$, $\zeta \neq 0$, from p(s) .

c) $\pi(s+1)$ " " " T-2, from p(s) .

d) $\pi(s+1)$ monoidal with center (x(s),z(s)) .

e) $\pi(s+1)$ " " " (y(s),z(s)).

(3.1.3) Let us denote by $\beta(t) = \beta(\Delta(\mathcal{D}(t),E(t),p(t))$. Assume that the player A has not won in stat (s+1). This section is devoted to prove the following theorem.

<u>Theorem</u>. a) If $\pi(s+1)$ is quadratic and $\beta(0) \leq 1+1/r$ then the player A can always win.

b) If $\pi(s+1)$ is quadratic and $\beta(0) > 1+1/r$, then the player A can obtain a bridge type (or win).

c) If $\pi(s+1)$ is monoidal with center (x(s),z(s)), then the player A can obtain a bridge type (or win).

d) If $\pi(s+1)$ is monoidal with center (y(s),z(s)), then the player

A can always win.

(3.2) The transformation $T_{-1,\zeta}$ $\zeta \neq 0$

(3.2.1) In this paragraph we shall prove the following

Theorem. If $r \geq 3$, then the player A always wins in this situation. If $r = 2$ and $\beta(0) = \beta(\Delta(\mathcal{D}(0), E(0), p(0))) \leq 1+1/r$ then the player A always wins. Otherwise, the player A wins or obtain a bridge type.

As a corollary of this, one obtains the corresponding part of the proof of the theorem (3.1.3)

(3.2.2) Lemma. Without loss of generality , one can assume that p(s) satisfies that if $\mathcal{D}(s)$ is generatd by

(3.2.2.1) $$D(s) = a(s)x(s)\,\partial/\partial x(s) + b(s)y(s)\,\partial/\partial y(s) + c(s)\,\partial/\partial z(s)$$

then

(3.2.2.2) $$(h,i,j) \in \mathrm{Exp}\,(b(s)-a(s);p(s)) \Rightarrow j \neq r-1.$$

Proof. The same proof of (1.2.5).

(3.2.3) Lemma. $(0,0,r) \in \mathrm{Exp}\,(b(s);p(s))$ and also $(0,0,r) \in \mathrm{Exp}\,(b(s)-a(s);p(s))$.

Proof. It follows from the proof of (chapter III.(2.1.3)).

(3.2.4) Assume that $\mathcal{D}(s+1)$ is generated by

(3.2.4.1) $$D(s+1) = a(s+1)x(s+1)\,\partial/\partial x(s+1) + b(s+1)\,\partial/\partial y(s+1) + c(s+1)\,\partial/\partial z(s+1)$$

with

(3.2.4.2) $$a(s+1) = a(s)/x(s+1)^r$$
$$b(s+1) = (y(s+1)-\zeta)(b(s)-a(s))/x(s+1)^r$$
$$c(s+1) = c(s)/x(s+1)^{r+1} - z(s+1)a(s)/x(s+1)^r.$$

Then, since $\pi(s+1)$ is nor standard nor natural one has

(3.2.4.3) $$\nu(c(s+1)) = r.$$

First let us observe that in view of (3.2.2) and (3.2.3) one has

(3.2.4.4) $$J^r(b(s+1)) \ni \underline{z}(s+1).$$

So if the player A has not won in this movement, one can assume that

(3.2.4.5) $$J^r(c(s+1)) = (\underline{x}(s+1))$$

since otherwise type zero or better. (By the way, this implies that

(3.2.4.6) $$J^r(b(s+1)) = (\underline{z}(s+1) + \lambda\underline{x}(s+1)) \quad \text{or} \quad (\underline{x}(s+1),\underline{z}(s+1))).$$

If $(x(s+1),z(s+1))$ is permissible the player A wins by choosing this center by (3.2.4.5) and (3.2.3). Assume that it is not permissible. If the player A chooses the quadratic center, then the player B must choose T-2 from p(s+1). Assume that $\mathcal{D}(s+2)$ is generated by

(3.2.4.7) $$D(s+2) = a(s+2)x(s+2)\partial/\partial x(s+2) + b(s+2)y(s+2)\partial/\partial y(s+2) + \\ + c(s+2)\partial/\partial z(s+2)$$

If the player A has not won, then one has type I' (in fact a bridge type). Let $p'(s+2) = (x'(s+2),y'(s+2),z'(s+2)) = (y(s+2),z(s+2),x(s+2))$. If

(3.2.4.7) $$\underline{\nu}_{\{2,3\}}(\mathcal{D}(s+2),E(s+2);p'(s+2)) \geq 2$$

then the player A wins as in (2.3.7) by choosing the center $(x(s+1),z(s+1))$. Thus assume that

(3.2.4.8) $$\underline{\nu}_{\{2,3\}}(\mathcal{D}(s+2),E(s+2);p'(s+2)) \leq 1.$$

(3.2.5) The situation will be reduced to a situation which has been already studied in (2.1) and (2.2). Looking at (3.2.4.2) and in view of (3.2.4.5) and (3.2.4.6) one has that

(3.2.5.1) $$a(s+1) = \phi(y(s+1), z(s+1)+x(s+1)(\ldots)$$
$$b(s+1) = \lambda z(s+1)^r + x(s+1)(\ldots), \quad \lambda \neq 0$$
$$c(s+1) = \Phi(y(s+1), z(s+1)+x(s+1)(\ldots)$$

where ϕ and Φ are homogeneus of degrees r and r+1 respectively. Let us make $p'(s+2) = (x',y',z')$ in order to simplify notation. One has that

(3.2.5.2) $$D(s+2) = x'(\Phi(1,y')\partial/\partial y' + \phi(1,y')z'\partial/\partial z') +$$
$$+\lambda y'^r(x'\partial/\partial x' - y'\partial/\partial y' - z'\partial/\partial z') + z'D^*(s+2).$$

Since the adapted order has not dropped, one has that

(3.2.5.3) $$\Phi(1,y) = \alpha y^{r+1} + \beta y^r + \gamma y^{r-1}$$
$$\phi(1,y) = \delta y^r + \varepsilon y^{r-1}$$

Moreover, if $\gamma \neq 0$, one deduces easily that

(3.2.5.4) $$J^r(c(s+2)) = (\underline{x}', \underline{y}', \underline{z}')$$

(remark that $c(s+2)$ is the coefficient of $\partial/\partial y'$) and then the directrix has dimension zero.

(3.2.6) Proposition. With the above notations, if $(\beta, \varepsilon) \neq (0,0)$, then the player A wins by choosing always the quadratic center.

Proof. Let us prove that the player B must choose always the transformation T-1,0. If the player A does not win this contradicts the fact (3.2.4.8) which implies that $(y'(s+2), z'(s+2))$ is not permissible. If the player B has chosen t times the transformation T-1,0 (and the adapted order has not dropped) one has

(3.2.6.1) $$D(s+2+t) = x''(\beta y''^r \partial/\partial y'' + \varepsilon y''^{r-1} z''\partial/\partial z'') +$$
$$+\lambda x''^t y''^r (x''\partial/\partial x'' - (t+1-(\alpha/\lambda)x''^{t+1})y''\partial/\partial y'' -$$
$$-(t+1-(\delta/\lambda)x''^{t+1})z''\partial/\partial z'') + z''D^*(s+2+t)$$

where $p'(s+2+t) = (x'',y'',z'')$ in order to simplify the notation. Obviously

(3.2.6.2) $$\underline{z}'' \in J(\mathcal{D}(s+2+t), E(s+2+t))$$

and the next quadratic transformation must be given by $(T-1,0)$, $(T-1,\zeta)$ or $T-2$. An easy computation over (3.2.6.1) shows that if the player B chooses $(T-1,\zeta)$, $\zeta \neq 0$ or $T-2$, then the adapted order drops.

(3.2.7) Proposition. With the hypothesis of (3.2.4), if the player A does not win, then the player B must choose at least $r-1$ times the transformation $(T-1,0)$ from $p'(s+2)$.

Proof. By (3.2.6) one can assume $\beta = \varepsilon = 0$ and to apply the proof of (2.2.7) to

(3.2.7.1)
$$D(s+2) = \lambda y'^{r}(x'\partial/\partial x' - (1-(\alpha/\lambda)x')y'\partial/\partial y' - $$
$$-(1-(\delta/\lambda)x')z'\partial \quad z') + z'D^*(s+2).$$

(3.2.8) Let us denote $y_1(s) = y(s)+\zeta x(s)$ and $p_1(s) = (x(s),y_1(s),z(s))$. Let us denote

(3.2.8.1)
$$\mathrm{Exp}^* = \mathrm{Exp}\ (y_1(s)a(s)/x(s);p_1(s)) \cup$$
$$\cup\ \mathrm{Exp}\ ((y_1(s)-\zeta x(s))(b(s)-a(s)/x(s));p_1(s)) \cup$$
$$\cup \mathrm{Exp}\ (y_1(s)(c(s)-z(s)a(s))/x(s)z(s);p_1(s)).$$

Looking at (3.2.4.2), there is a bijection

(3.2.8.2)
$$\phi\colon \mathrm{Exp}^* \longrightarrow \mathrm{Exp}\ (D(s+2),E(s+2);p'(s+2))$$

given by

(3.2.8.3)
$$\phi\ (h,i,j) = (h+2(i+j-r),j+i,h+i+j-r)$$

(compare with (2.3.5)).

(3.2.9) Proposition. With the hypothesis of (3.2.4), if $r \geq 3$ then the player A wins in less than $r-1$ movements by choosing always the quadratic center.

Proof. One can reason as in (2.3.6). Since now $-1 \leq r,j$, from (2.3.6.9) we obtain $i-r \leq r+1$ and thus $h' \leq r+1$. But (2.3.6.11) is satisfied with this condition also. Now it is enough to apply the above proposition as in (2.3.8) a).

(3.2.10) Proposition. With the hypothesis of (3.2.4), if $\beta(0) \leq 1+1/r$ and $r = 2$, then the player A will win in less that $r+1 = 3$ movements by choosing always the quadratic center.

Proof. By reasoning as in (2.3.8) b), it is enough to prove that there is a vertex $w = (u,v) \in \Delta(s+2) = \Delta(D(s+2),E(s+2);p'(s+2))$ such that

$$(3.2.10.1) \qquad u + 2v \leq 2$$

(if $v = 1$ one has type zero as in the final part of the proof of (2.3.6)).

First, since $\beta(0) \leq 1+1/r$ and $\pi(1)$ is given by T-2 or T-4, one deduces easily that

$$(3.2.10.2) \qquad \beta(s) \leq \beta(1) \leq 1$$

(remark that a vertex of $\Delta(0)$ is given by $(h/(r-j),i/(r-j))$ for some (h,i,j) and thus it may never be $(1-1/(r+1),1+1/r)$: only possibility in order to make (3.2.10.2) false).

Assume that (3.2.10.1) is not true. Then, by looking at the proof of (2.3.6) and the definition of Exp* in (3.2.8.1), the only possibility for finding $(h',i',j') \in \text{Exp}\,(D(s+2),E(s+2);p'(s+2))$ such that $i'+j' \leq 1$ and

$$(3.2.10.3) \qquad h'/(2-j') + 2i'/(2-j') \leq 3$$

is that

$$(3.2.10.4) \qquad (h,i,j) = \phi^{-1}(h',i',j') = (-1,5,-1)$$

(where ϕ is as in (3.2.8.2)). But this implies easily that

$$(3.2.10.5) \qquad (0,4,0) \in \text{Exp}\,(c(s);p(s))$$

Then the main vertex $(\alpha(s),\beta(s))$ of $\Delta(s)$ has $\alpha(s)=0$. So $\beta(s)=1$, and $(\alpha(s),\beta(s))= =(0,1)$. This is a contradiction in the following way: by (3.2.10.2), $\beta(1) = 1$ and since $\Delta(1) = \sigma(\Delta(0))$, with $\sigma(u,v) = (u,u+v-1)$, necessarily $\alpha(1) = ½$ or 1/3 (i.e. $1/r$ or $1/(r+1)$); since the ordinate of the main vertex is always one, the transformations $\pi(t)$, $t=2,\dots,s$ are necessarily given by T-1,0 and in this way it is not possible that the main vertex of $\Delta(s)$ would be (0,1).

(3.2.11) The above proposition ends the proof of the theorem (3.2.1), since D(s+2) is of the bridge type (see (3.2.4.7)).

(3.3) The transformation T-2

(3.3.1) Here the corresponding part of the theorem (3.1.3) will be proved.

(3.3.2) Assume that $\mathcal{D}(s+1)$ is generated by

(3.3.2.1) $$D(s+1)=a(s+1)x(s+1)\partial/\partial x(s+1)+b(s+1)y(s+1)\partial/\partial y(s+1)+c(s+1)\partial/\partial z(s+1)$$

with $a(s+1) = (a(s)-b(s))/y(s+1)^r$, $b(s+1) = b(s)/y(s+1)^r$, $c(s+1) = c(s)/y(s+1)^{r+1} - z(s+1)b(s)/y(s+1)^r$. One can assume that $\nu(c(s+1)) = r$. If the player A has not won, then one has the following possibilities for $J^r(c(s+1)) = J(\mathcal{D}(s+1),E(s+1))$:

(3.3.2.2)
$$J^r(c(s+1)) = (\underline{x}(s+1))$$
$$J^r(c(s+1)) = (\underline{y}(s+1)+\lambda\underline{x}(s+1))$$
$$J^r(c(s+1)) = (\underline{x}(s+1),\underline{y}(s+1))$$

In the third case, the player A wins by the fact

(3.3.2.3) $$(0,0,r) \in \mathrm{Exp}\ (b(s+1);p(s+1))$$

which can be easily deduced from (3.2.3) or (III.(2.1.3)).

(3.3.3) Assume first that $J^r(c(s+1)) = (\underline{y}(s+1)+\lambda\underline{x}(s+1))$. This implies that

(3.3.3.1) $$(0,2r+1,-1) \in \mathrm{Exp}\ (\mathcal{D}(s),E(s);p(s))$$

and

(3.3.3.2) $$(0,2-1/(r+1)) \in \Delta(\mathcal{D}(s),E(s);p(s)) = \Delta(s)$$

This is the main vertex of the polygon $\Delta(s)$. Then

(3.3.3.3) $$\beta(s) = 2 - 1 / (r+1).$$

Now, since $\beta(0) > \beta(s)$ and one has a bridge type, by making

(3.3.3.4) $$p'(s+1) = (x'(s+1),y'(s+1),z'(s+1)) = (x(s+1),z(s+1),y(s+1))$$

then the part b) of the theorem (3.1.3) is proved in this case.

(3.3.4) Assume now that $J^r(c(s+1)) = (\underline{x}(s+1))$. Let us denote by

(3.3.4.1) $$f: \mathbb{R}^3 - \{(h,i,j); j \geq r\} \longrightarrow \mathbb{R}^2$$

the projection given by $f(h,i,j) = (h/(r-j),i/(r-j))$. Since the only monomial of degree r in $c(s+1)$ is $x(s+1)^r$, one has that

(3.3.4.2) $$(r,1,-1) \in \text{Exp } (D(s),E(s);p(s)) = \text{Exp } (s)$$

and that

(3.3.4.3) $$f(\text{Exp } (s) - \{(r,1,-1)\}) \subset \{(u,v);\ 2u+v \geq 2\}.$$

Now, if $\beta(0) > 1+1/r$ one has a type bridge as above, by making $p'(s+1) = (x'(s+1),y'(s+1),z'(s+1)) = (y(s+1),z(s+1),x(s+1))$. Assume that $\beta(0) \leq 1+1/r$. Since $\pi(1)$ is given by T-2 or T-4 (actually by T-2) one has that

(3.3.4.4) $$\beta(s) \leq \beta(1) < \beta(0) \leq 1+1/r$$

This implies that $\beta(s) \leq 1$, since the main vertex of $\Delta(0)$ cannot be $(1-1/(r+1),1+1/r)$, which is the only obstruction to prove the last assertion, let us observe that the vertex must be of the form $(h/(r-j),i/r-j))$ for a certain (h,i,j)).

By combining (3.3.4.3) and the fact $\beta(s) \leq 1$, one deduces that if $(\alpha(s),\beta(s))$ is the main vertex of $\Delta(s)$, then

(3.3.4.5) $$\alpha(s) \geq 1/2.$$

Let us now consider the bijection

(3.3.4.6) $$\phi : \text{Exp } (s) \longrightarrow \text{Exp } (D(s+1),E(s+1);p'(s+1))$$

which is given by

(3.3.4.7) $$\phi(h,i,j) = (h+i+j-r, j+1, h)$$

Now, let $(h',i',j') \in \mathrm{Exp}\,(D(s+1),E(s+1);p'(s+1))$ and let us suppose that $\phi^{-1}(h',i',j') = (h,i,j)$. By (3.3.4.5) one has that

(3.3.4.8) $$2h + j \geq r \Rightarrow 2j'+i' \geq r+1 \Rightarrow$$
$$\Rightarrow j'+i' \geq j'+i'/2 \geq (r+1)/2 > 1$$

Then $j'+i' \geq 2$ and

(3.3.4.9) $$\underline{\nu}_{\{2,3\}}(\mathcal{D}(s+1),E(s+1);p'(s+1)) \geq 2.$$

By (3.3.2.3) if the player A chooses the center $(y'(s+1),z'(s+1))$ then he always wins.

Thus, the theorem (3.1.3) is proved in this case.

(3.4) The transformation T-1,0.

(3.4.1) In this paragraph the corresponding part of the proof of the theorem (3.1.3) will be made.

(3.4.2) Assume that $\mathcal{D}(s+1)$ is generated by

(3.4.2.1) $$D(s+1) = a(s+1)x(s+1)\partial/\partial x(s+1) + $$
$$+ b(s+1)y(s+1)\partial/\partial y(s+1) + c(s+1)\partial/\partial z(s+1)$$

with $a(s+1) = a(s)/x(s+1)^r$, $b(s+1) = (b(s)-a(s))/x(s+1)^r$, $c(s+1) = c(s)/x(s+1)^{r+1} - z(s+1)a(s+1)$, and $\nu(c(s+1)) = r$. If the player A has not won, one has the following possibilities

(3.4.2.2) $$J^r(c(s+1)) = (\underline{x}(s+1))$$
$$J^r(c(s+1)) = (\underline{y}(s+1)+\lambda\,\underline{x}(s+1))$$
$$J^r(c(s+1)) = (\underline{x}(s+1),\underline{y}(s+1))$$

In the third case, the player A wins since

(3.4.2.3) $(0,0,r) \in \mathrm{Exp}\,(b(s+1);p(s+1))$

which can be deduced from (3.2.3) or (III. (2.1.3)).

(3.4.3) Proposition. Assume that $\beta(0) \leq 1+1/r$ and that the player A has not won. Then no one of the transformations $\pi(i)$, $1 \leq i \leq s$ is given by T-4.

Proof. If any (i) is given by T-4, then

(3.4.3.1) $$\beta(s) \leq \beta(0)-1 \leq 1/r$$

Since $\pi(s+1)$ is not T-3, the main vertex of $\Delta(s)$ is $(1-1/r,1/r)$ or $(1-1/(r+1),1/(r+1))$ and in both cases the adapted order drops with T-1,0.

(3.4.4) Assume that $J^r(c(s+1)) = (\underline{x}(s+1))$, if $\beta(0) > 1+1/r$ and the player A has not won, one has the bridge type by putting $p'(s+1) = (x'(s+1),y'(s+1),z'(s+1)) = (y(s+1),z(s+1),x(s+1))$.

Assume that $\beta(0) \leq 1+1/r$. First, one has

(3.4.4.1) $$(2r+1,0,-1) \in \mathrm{Exp}\,(s)$$

(3.4.4.2) $$(2-1/(r+1),0) \in \Delta\,(s).$$

On the other hand, by (3.4.3),

(3.4.4.3) $$\Delta(t) = \sigma_t(\Delta(t-1))$$

for $t=1,\ldots,s$, where $\sigma_1(u,v) = (u,v+v-1)$ and for $t \geq 2$,

(3.4.4.4) $$\sigma_t(u,v) = (u+v-1,v)$$

or $$\sigma_t(u,v) = (u,u+v-1)$$

or $$\sigma_t(u,v) = (u-1,v)$$

and this contradicts (3.4.4.2) (may be proved by induction since $(1,0) \notin \Delta(0)$). Thus the theorem (3.1.3) is proved in this case for $J^r(c(s+1)) = (\underline{x}(s+1))$.

(3.4.5) Proposition. If $\beta(0) > 1+1/r$, then the player A can always win or obtain the bridge type.

Proof. In the case $J^r(c(s+1)) = (\underline{y}(s+1)+\lambda\underline{x}(s+1))$, a change $p'(s+1) = (x'(s+1),y'(s+1),z'(s+1)) = (x(s+1),z(s+1),y(s+1))$ allow us to see easily that we have the bridge type (join to (3.4.2.3)), if the player A has not won.

(3.4.6) Assume now that $J^r(c(s+1)) = (\underline{y}(s+1)+\lambda\underline{x}(s+1))$. Then

(3.4.6.1) $$(1,r,-1) \in \text{Exp}(s)$$

(3.4.6.2) $$(1/(r+1),1-1/(r+1)) \in \Delta(s).$$

This implies that

(3.4.6.3) $$\beta(s) \geq 1 - 1/(r+1)$$

(3.4.7) Lemma. If $\beta(0) \leq 1+1/r$ and $J^r(c(s+1)) = (\underline{y}(s+1)+\lambda\underline{x}(s+1))$, then $\pi(i)$ is T-2 or T-1,0 for each i, $1 \leq i \leq s$. Moreover, let t be the biggest index such that $\pi(t)$ is given by T-2, then

(3.4.7.1) $$s-t \geq r-1.$$

Proof. Let $(\alpha(0),\beta(0))$ be the main vertex of $\Delta(0)$. Since $\beta(s) \leq \beta(1)$, $\pi(1)$ is given by T-2, $\alpha(0) < 1$ and $\beta(0) \leq 1+1/r$, there are only three possibilities for the main vertex of $\Delta(0)$:

(3.4.7.2) $$(\alpha(0),\beta(0)) = (1-1/r,1+1/r)$$
$$(1-1/(r+1),1)$$
$$(1-1/(r+1),1+1/(r+1)).$$

then, the main vertex of $\Delta(1)$ verifies

(3.4.7.3) $$(\alpha(1),\beta(1)) = (1-1/r,1) \quad \text{or}$$
$$(1-1/(r+1),1-1/(r+1)) \quad \text{or}$$
$$(1-1/(r+1),1)$$

Now T-3 will never be applied, this proves the first part. For the second part, let us distinguish two cases: t=1 or t >1. If t=1

(3.4.7.4) $$\Delta(s) = \sigma^{s-1}(\Delta(1))$$

where $\sigma(u,v) = (u+v-1,v)$ and the result follows from (3.4.6.2) and the fact that if (α,β) is a vertex of $\Delta(1)$ with $\beta = 1-1/(r+1)$, then $\alpha \geq 1-1/(r+1)$. If $t>1$, since $\beta(t-1) \leq 1$, then necessarily $\alpha(t-1) = 1-1/(r+1)$ (otherwise (3.4.6.2) is never reached) and so $(\alpha(t),\beta(t)) = (1-1/(r+1),1-1/(r+1))$, now, one can reason as above, since

(3.4.7.5) $$\Delta(s) = \sigma^{s-t}(\Delta(t)).$$

(3.4.8) <u>Lemma</u>. If $\beta(0) \leq 1+1/r$ and $J^r(c(s+1)) = (\underline{y}(s+1)+\lambda\underline{x}(s+1))$ then for each $(h,i,j) \in \mathrm{Exp}(s)$ one has that $i+j \geq 1$.

<u>Proof</u>. First, let us see that $\lambda = 0$. If $\lambda \neq 0$, then $(2r+1,0,-1) \in \mathrm{Exp}(s)$ and thus $(2-1/(r+1),0) \in \Delta(s)$ and it is a vertex of $\Delta(s)$ (otherwise the initial form of c(s+1) will be different or the adapted order will drop). Let us take the notations of (3.4.7), then this vertex comes from a vertex (α,β) of $\Delta(t-1)$, where $\alpha>1$, but this is impossible since $\pi(t)$ is given by T-2, and one cannot obtain the ordinate zero.

Let $f\colon \mathbb{R}^3 - \{(h,i,j)\,;\, j\geq r\} \longrightarrow \mathbb{R}^2$ be as in (3.3.4.1), since $\lambda = 0$, reasonning as in (3.3.4), one has that

(3.4.8.1) $$f(\mathrm{Exp}(s) - \{(1,r,-1)\}) \subset \{(u,v); u+2v \geq 2\}.$$

Let $(h,i,j) \in \mathrm{Exp}(s)$. If $j \geq r$ or $(h,i,j) = (1,r,-1)$ one has that $i + j \geq 1$. Let

(3.4.8.2) $$(\alpha,\beta) = (h/(r-j),i/(r-j)) \in \Delta(s).$$

then $\alpha + 2\beta \geq 2$. Let us denote

(3.4.8.3) $$\sigma(u,v) = (u+v-1,v)$$
$$\sigma'(u,v) = (u,u+v-1).$$

By the lemma (3.4.7), there is $(\alpha',\beta') \in \Delta(t-1)$ such that

(3.4.8.4) $$\sigma^{s-t}(\sigma'(\alpha',\beta')) = (\alpha,\beta)$$

and $s-t \geq r-1$. Let $(\alpha'',\beta'') = \sigma'(\alpha',\beta')$, then

(3.4.8.5) $$(\alpha,\beta) = (\alpha''+(s-t)(\beta''-1),\beta'')$$

and since $\alpha+2\beta \geq 2$,

(3.4.8.6) $$\alpha'' + (r+1)\beta'' \geq r+1.$$

Now (3.4.8.6) implies

(3.4.8.7) $$\text{if } \alpha'' \leq |(r+1)/(r+2)|.2 \Rightarrow \beta'' \geq r/(r+2)$$

If $\alpha'' = \alpha' \geq |(r+1)/(r+2)|.2$ then

(3.4.8.8) $$\beta'' = \beta' + \alpha'-1 \geq \frac{r+1}{r+2} . 2 - 1 = \frac{r}{r+2}$$

And since $r \geq 2$,

(3.4.8.9) $$i/(r-j) = \beta = \beta'' \geq 1/2 \Rightarrow 2i+j \geq r$$

If $i+j \leq 0$ then $(i,j) = (1,-1)$ or $(0,-1)$ or $(0,0)$ and never $2i+j \geq r$, contradiction.

(3.4.9) <u>Proposition</u>. If $\beta(0) \leq 1+1/r$ and $J^r(c(s+1)) = (\underline{y}(s+1)+\lambda\underline{x}(s+1))$ then the player A wins by choosing the monoidal center $(y(s+1),z(s+1))$.

<u>Proof</u>. Let $p'(s+1) = (x'(s+1),y'(s+1),z'(s+1)) = (x(s+1),z(s+1),y(s+1))$. Then one has the bridge type, unless A has won, and it is enough to prove that

(3.4.9.1) $$\underline{\nu}_{\{2,3\}}(D(s+1),E(s+1);p'(s+1)) \geq 2.$$

There is a bijection

(3.4.9.2) $$\phi\colon \operatorname{Exp}(s) \longrightarrow \operatorname{Exp}(D(s+1),E(s+1);p'(s+1))$$

given by $\phi(h,i,j) = (h+i+j-r,j+1,i)$. Then (3.4.9.1) follows immediatly from the lemma (3.4.8)

(3.5) $\pi(s+1)$ monoidal with center $(x(s),z(s))$

(3.5.1) If the player A does not win he will obtain always a bridge type.

(3.5.2) Since $J(\mathcal{D}(s),E(s)) \ni \underline{z}(s)$, one can suppose that $\pi(s+1)$ is given by T-3 from $p(s)$ and if $\mathcal{D}(s+1)$ is generated by

(3.5.2.1) $$D(s+1) = a(s+1)x(s+1)\partial/\partial x(s+1) + b(s+1)y(s+1)\partial/\partial y(s+1) + + c(s+1)\partial/\partial z(s+1)$$

where $a(s+1) = a(s)/x(s)^r$, $b(s+1) = b(s)/x(s)^r$, $c(s+1) = c(s)/x(s)^r - z(s+1)a(s+1)$, then $\nu(c(s+1)) = r$. If the player A has not won, one can also suppose that $J^r(c(s+1)) = J(\mathcal{D}(s+1),E(s+1))$ satisfies

(3.5.2.2) $$J^r(c(s+1)) = (\underline{x}(s+1))$$
$$J^r(c(s+1)) = (\underline{y}(s+1)+\lambda\underline{x}(s+1))$$
$$J^r(c(s+1)) = (\underline{x}(s+1),\underline{y}(s+1))$$

(The other possibilities produce type zero or dim Dir $(\mathcal{D}(s+1)),E(s+1)) = 0$). In any case, one has that

(3.5.2.3) $$(0,0,r) \in \mathrm{Exp}\ (b(s+1);p(s+1))$$

by the same reason as (3.2.3) or (III.(2.1.3)). This implies that in the third case of (3.5.2.2) the player A will win by choosing the quadratic center.

In the first and second of (3.5.2.2) one can see that $(X(s+1),E(s+1),\mathcal{D}(s+1),P(s+1))$ is of the bridge type in view of (3.5.2.3).

(3.6) $\pi(s+1)$ monoidal with center $(y(s),z(s))$

(3.6.1) The player A can always win in this case. The proof of this result is very similar to the proof of (2.6.1).

(3.6.2) Since $\underline{z}(s) \in J(\mathcal{D}(s),E(s))$ one can suppose that $\pi(s+1)$ is given by T-4 from p(s). We obtain D(s+1) as in (3.5.2.1) with $a(s+1)=a(s)/y(s+1)^r$, $b(s+1) = b(s)/y(s+1)^r$, $c(s+1) = c(s)/y(s+1)^r - z(s+1)b(s+1)$. Since the transition is not standard, $\nu(c(s+1)) = r$. We have also the possibilities of (3.5.2.2) and the property (3.5.2.3). In the third possibility the player A wins as in (3.5.2).

(3.6.3) Assume $J^r(c(s+1)) = (\underline{y}(s+1)+\lambda\underline{x}(s+1))$, this implies that

(3.6.3.1) $$(0,2r+1,-1) \in \text{Exp}\ (D(s),E(s);p(S)) = \text{Exp}\ (s)$$

and necessarily (0,2-1/(r+1)) is the main vertex of $\Delta(s)$, thus it is the "only" vertex (in particular $\lambda= 0$). Then one deduces that for each $(h,i,j) \in \text{Exp}\ (s)$, one has that

(3.6.3.2) $$i/(r-j) \geq 2-1/(r-1) \quad (\text{if } j < r)$$

and then one deduces that if I = (y(s+1),z(s+1)) then

(3.6.3.3) $$\nu_I(a(s+1)) \geq r; \quad \nu_I(b(s+1)) \geq r; \quad \nu_I(c(s+1)) \geq r$$

and the curve given by I is permissible. By the property (3.5.2.3), the player A wins by choosing this center.

(3.6.4) Assume $J^r(c(s+1)) = (\underline{x}(s+1))$. Then

(3.6.4.1) $$(r,r+1,-1) \in \text{Exp}(s)$$

and (1-1/(r+1),1) is the only vertex of $\Delta(s)$. One deduces that (x(s+1),z(s+1)) is permissible and the player A wins by choosing this center. (See also (2.6)).

(3.6.5) This ends the proof of the theorem (3.1.3)

4. A WINNING STRATEGY FOR THE TYPE ONE

(4.1) Introduction

In this section we shall establish a winning strategy for the player A when the reduction game begins with any type one.

All the computations which have been made in this chapter and in the chapter III allow us to state the following:

(4.1.1) Theorem. If the reduction game begins with any type one I-2 or I-1-0, then there is a strategy for the player A in order to obtain the victory or to obtain a bridge type. And if it begins with type I-1-1 or I'-1-1, there is a strategy in order to obtain the victory or to obtain a type I'-2.

It remains to study the game when it begins by a type one I'-1-0 or I'-2, or by the bridge type. In this section we shall prove the following two theorems:

(4.1.2) Theorem. If the reduction game begins with any type one I'-1-0 or I'-2, then there is a strategy for the player A in order to obtain the victory or to obtain a bridge type.

(4.1.3) Theorem. If the reduction game begins with a bridge type, then there is a winning strategy for the player A.

(4.2) Standard transitions from the type I'

(4.2.1) Let $(X,E,\mathcal{D},P)$ be of the type I'-2 or I'-1-0. In this paragraph, the existence of a winning strategy for the player A for the "standard" transitions will be proved.

(4.2.2) Definition. Let $p = (x,y,z)$ be a normalized base. p is called I'-prepared iff there is no change $y_1 = y + \zeta x^n$ such that if $p_1 = (x,y_1,z)$ then $(\varepsilon(\Delta),-l(\Delta)) < (\varepsilon(\Delta_1),-l(\Delta_1))$ for the lexicographic order, where

$$\Delta = \Delta(\mathcal{D},E,p);\ \Delta_1 = \Delta(\mathcal{D},E,p_1).$$

From p one can obtain always an I'-prepared base by a sequence of changes $y_1=y+\zeta x^n$.

(4.2.3) Definition. Let G be a realization of the reduction game beginning at $(X,E,\mathcal{D},P)$

and let p = (x,y,z) be a normalized base:

a) G is "standard" until the step $s <$ length (G) iff $P(t) \in$ strict transform of (z=0) for t=0,1,...,s.

b) Assume that p is I'-prepared. Then G follows the "standard winning strategy" until the step $s <$ length (G) iff G is standard until the step s and for $0 \leq t \leq s$, there is a regular system of parameters p(t) such that

. b-1. p(0) = p

. b-2. If e(E(t)) = 3, then p(t) is obtained from p(t-1) by (T-1,0),T-2, T-3 or T-4. If e(E(t)) = 2 (hence stat (t) is of the type I') then there is p'(t) obtained from p(t-1) by $(T\text{-}1,\zeta)$, T-2, T-3 or T-4 and p(t) is I'-prepared obtained from p'(t).

. b-3. The center Y(t), $0 \leq t \leq s$, verifies the usual assumptions for the 0-retarded standard strategy with respect to p(t) (see III.2.8.4).

(4.2.8) Remarks. Let us assume tacitly that type zero is considered as a victory situation , then,in (4.2.3) one has that stat (t), $0 \leq t \leq s$ is always of the types I'-2, I'-1-0, II'-2, or II'-1. (see III. (2.4.1)). On the other hand, there is a strategy for the player A in such a way that each realization of the game G which follows the strategy verifies that if G is standard until the step s then G follows the standard winning strategy until the step s.

(4.2.5) Theorem. Let G be a realization of the game suchthat for each $s <$ length (G) then G follows the standard winning strategy. Then G is finite.

Proof. If the transitions are of the type $I' \mapsto II'$ or $II' \mapsto II'$, then the result follows from the usual control of the polygon (see the standard transition of the chapter III). It remains to control the transition $II' \longmapsto I'$, but this is similar to the "natural" transition of the section 1 (actually, it is slightly easier since there is no good preparations $z \mapsto z_1$).

(4.3) No standard transitions from the type I'

(4.3.1) Here the proof of the theorem (4.1.2) will be finished.

(4.3.2) Proposition. Let $(X,E,\mathcal{D},P)$ be of the type I'-1-0 or I'-2 and let $p=(x,y,z)$ be an I'-prepared base. Let G be a realization of the reduction game beginning at $(X,E,\mathcal{D},P)$ such that G follows the standard winning strategy until the step s-1 and G is not standard until the step $s <$ length (G) (with respect to p). Then:

a) If E_1 is given by $x = 0$, then $\mathcal{D} = (\mathcal{D},E_1)$ and $\mathcal{D}$ is multiplicatively irreducible and adapted to E_1.

b) Let G' be obtained by putting mov'(t) = mov(t) and stat'(t) be obtained inductively from stat'(0) = $(X,E_1,\mathcal{D},P)$. Then G' is a realization of the reduction game of length $\geq$ s+1 and stat'(s) = stat(s). Moreover for $t=0,1,\ldots,s-1$

(4.3.2.1) stat(t) is of the type I' (resp. II') $\Leftrightarrow$

$\Leftrightarrow$ stat'(t) is of the type I (resp II).

Proof. Assume that

(4.3.2.2) $$D(t) = a(t)x(t)\partial/\partial x(t) + b(t)\partial_t + c(t)z(t)\partial/\partial z(t)$$

where $p(t) = (x(t),y(t),z(t))$ is obtained as in (4.2.3) and $\partial_t = \partial/\partial y(t)$ or $y(t)\partial/\partial y(t)$, depending on $e(E(t))$. By arguments as in (III. (2.1.3)) one has that $\nu(b(t)) = r$, $0 \leq t \leq s-1$ and that $z(t)$ divides simultaneously $a(t)$ and $b(t)$ iff $z(t-1)$ divides simultaneausly $a(t-1)$ and $b(t-1)$, $t=1,\ldots,s-1$. If $z(s-1)$ divides $a(s-1)$ and $b(s-1)$, then $P(s) \in$ strict transform of $z = 0$ (otherwise the adapted order drops), then, for $t=0,\ldots,s-1$, $z(t)$ does not divide $a(t)$ and $b(t)$. If $E_2(t)$ is the strict transform of $z = 0$ and $E(t) = E'(t) \cup E_2(t)$, from the above result one deduces that $\mathcal{D}(t)$ is multiplicatively irreducible and adapted to $E'(t)$. In particular a) is proved.

Let $f(t) = \pi(t) \ \ldots \ \pi(1)$, then

(4.3.2.3)
$$\text{stat}(t) = (X(t),E(t),(\alpha(\mathcal{D}^{f(t)}),E(t)),P(t))$$
$$\text{stat}'(t) = (X(t),E'(t),(\alpha(\mathcal{D}^{f(t)}),E'(t)),P(t))$$

(see I.(1.2) and I.(1.3)). But since $\mathcal{D}(t)$ is multiplicatively irreducible and adapted to E'(t), one has

(4.3.2.4)
$$(\alpha(\mathcal{D}^{f(t)}),E'(t)) = (\alpha(\mathcal{D}(t)),E'(t)) = \mathcal{D}(t)$$

t=0,1,...,s. The proof of b) is finished by remarking that $\nu(\mathcal{D}(t),E'(t))=r$, t=0,1.. ..,s, since $\nu(b(t)) = r$. (the other statements follows straigthforward).

(4.3.3) Remarks. From the above proof one deduces that the centers Y(t), t=0,1,... ...,s-1 are always tangents to Dir $(\mathcal{D}(t),E(t))$ = Dir $(\mathcal{D}(t),E'(t))$. Moreover, by looking at (III.(2.4.1)) one obtains that stat'(t) is of the type I-1-0, I-2, II-2, II-1-2-0 or II-1-1-0.

(4.3.4) Definition. G' above is called the I-equivalent realization of G with respect to p.

(4.3.5) Theorem. Let $(X,E,\mathcal{D},P)$ be of the type I-1-0 or I-2 and let G be a realization of the reduction game such that $\pi(t)$ is standard for t=1,...,s-1, Y(t) is permissible and tangent to the directrix for t=0,...,s-1 and $\pi(s)$ is not standard ($s \leq$ length (G)). Then stat(s) is a victory situation, or of the type I-1-0 or I-2, or there is a strategy for the player A in order to obtain the bridge type.

Proof. Assume that stat(s) is not a victory stuation, nor of the type I-1-0 nor I-2. Two cases: stat(s-1) is of the type I (i.e. I-1-0 or I-2) or stat(s-1) is of the type II (i.e. II-2 or I-1-1-0 or II-1-2-0).

In the first case one can assume s=1. If $\pi(1)$ is quadratic, let us fix p=(x,y,z) well prepared. Then one can apply the computations of (2.2), (2.4) in order to obtain a bridge type. If $\pi(1)$ is monoidal with center Y(0) contained in E(0) then one can choice p such I(Y(0)) = (x,z) and the result follows from (2.5). If Y(0) is transversal to E(0), one can assume I(Y(0)) = (y,z) and the computations of (2.6) may be used.

In the second case, one can assume that stat(1) is of the type II. Let us prove that there is a strongly well prepared r.s. of p. p(s-1) such that $\mathcal{D}(s-1)$ is generated by

$$(4.3.5.1) \qquad D(s-1) = a(s-1)x(s-1)\partial/\partial x(s-1) + b(s-1)y(s-1)\partial/\partial y(s-1) + \\ + c(s-1)\partial/\partial z(s-1)$$

in such a way that there is $l \geq 0$, $m > 0$, $\lambda \neq 0$ with

$$(4.3.5.2) \qquad In^r(a(s-1)) = l.\lambda \underline{z}(s-1)^r \pmod{(\underline{x}(s-1),\underline{y}(s-1))}$$
$$In^r(b(s-1)) = m.\lambda \underline{z}(s-1)^r \pmod{(\underline{x}(s-1),\underline{y}(s-1))}$$

Let us proceed by induction: for the transition stat(0) ⟶ stat(1) one can choose p(0) well prepared in such a way that $\pi(1)$ is given by T-2 or T-4. For the transition stat(t) ⟶ stat(t+1) $t \geq 1$, $t+1 \leq s-1$, one can modify p(t) without touching the conditions (4.3.5.2) in such a way that if $\pi(t+1)$ in monoidal then it is given by T-3 or T-4 (if it is quadratic there is no problem) and the result follows straighforward. Now, the result follows from the computations of the section 3, since (4.3.5.2) is the only property that one needs to obtain a bridge type or to win.

(4.3.6) <u>(Proof of the theorem</u> (4.1.2)). Let us define a strategy in order to prove (4.1.2). Let us fix an I'-prepared base p = (x,y,z). Assume that $G|_t$ is a partial realization of the game which follows the strategy. One has to define the following center Y(t). If $G|_t$ is standard as in (4.2.3), then Y(t) is defined by the standard winning strategy of (4.2.3). Assume now that $G'|_t$ is standard until $t' < t$. Let $G'|_t$ be the I-equivalent realization, let us distinguish two possibilities:

a) $G'|_t$ is composed of standard or natural transitions. Then Y(t) is choiced in the usual way from $\Delta(\mathcal{D}(t),E(t),p(t))$, where p'(t) is obtained inductively from p(t') by (T-1,ζ), T-2, T-3 or T-4 and strong (very) good preparation.

b) No a). Then one applies the strategy of the theorem (4.3.5).

The above strategy is enough for proving the theorem (4.1.2). If one has b, then the result follows from the theorem (4.3.5), if one has a) or $G|_t$ is standard, the result follows from the usual control of the polygon.

(4.4) A winning strategy for the bridge type

(4.4.1) Definition. Let $(X,E,\mathcal{D},P)$ be of the type I-1-0 or I-2 and let $p = (x,y,z)$ be a very well prepared base. Let G be a realization of the reduction game beginning at $(X,E,\mathcal{D},P)$. G follows the "l-retarded general winning strategy" with respect to p iff

a) $\pi(t)$ is standard or natural $t < \text{length}(g)$

b) Assume that $\pi(t_1),\pi(t_2),\ldots,\ \ t_1 < t_2 < \ldots$ are the natural transitions, then $G|_{t_1}$ follows the l-retarded standard winning strategy with respect to p.

c) Let $p(t_1)$ be obtained as in the section 1, very well prepared, and let G_1 be obtained in a natural way from G as a realization beginning at stat (t_1). Then a), b) and c) are true if one begins with G_1, $p(t_1)$.

(Remark that one has has a recursive definition).

(4.4.2) Remark. The results of the chapter III and of the section 1 of this chapter show that if G follows the above strategy, then G is finite.

(4.4.3) Theorem. Let $(X,E,\mathcal{D},P)$ and $p = (x,y,z)$ be as in the above definition. Let us fix l. Let (α,β) be the main vertex of the polygon $\Delta(\mathcal{D},E,p)$ and assume that

(4.4.3.1) $$\beta \leq 1 + 1/r\ ,\ \alpha < 1.$$

Then, if G is any realization of the reduction game beginning at $(X,E,\mathcal{D},P)$ such that $G|_s$ follows the l-retarded general winning strategy and stat(s+1) is not a victory situation, then there is a winning strategy for the reduction game beginning at stat(s+1).

Proof. In view of (4.4.2), by induction, one can assume that $\Delta(s+1)$ is nor standard nor natural. If $\pi(s+1)$ is quadratic or it is centered at $(y(s),z(s))$, then the result follows from (3.1.3) and (2.1.3). Assume that $\pi(s+1)$ is centered at $(x(s),z(s))$. By (4.4.3.1) one can deduce that stat(t) is of the type I for $0 \leq t \leq s$, since otherwise one obtains the "stable situation" $\beta \leq 1$, $\alpha < 1$ (remark that after the natural transition, the main abscissa is preserved). Moreover, $\alpha(s-1) < 1$ and $\beta(s-1) > 1$, thus: $\beta(s-1) = 1 + 1/r$ or $1 + 1/r+1$. Now, $\pi(s)$ must be given by $(T-1,\zeta)$, ($\zeta = 0$ or not) and, since $(x(s),z(s))$ is permissible, one has that

(4.4.3.2) $$(\alpha(s-1),\beta(s-1)) = (1-1/r,1+1/r) \quad \text{or} \quad (1-1/(r+1),1+1/(r+1)).$$

This implies that

(4.4.3.3) $$(r,r+1,0) \in \operatorname{Exp}(D(s),E(s),p(s))$$

or $$(r+1,r+2,-1) \in \operatorname{Exp}(D(s),E(s),p(s)).$$

Assume that

(4.4.3.4) $$D(s+1) = a(s+1)x(s+1)\partial/\partial x(s+1) + b(s+1)\partial/\partial y(s+1) + c(s+1)\partial/\partial z(s+1)$$

Then from (4.4.3.3) on deduces that

(4.4.3.5) $$(0,r+1,0) \text{ or } (0,r+2,-1) \in \operatorname{Exp}(a(s+1)y(s+1)) \cup \operatorname{Exp}(b(s+1)) \cup \operatorname{Exp}(c(s+1)y(s+1)/z(s+1)).$$

Now, by (2.5.4), if the player A has not won in stat(s+1), then the following (quadratic) transformation must be T-2, and then the player A wins by (4.4.3.4).

(4.4.4) Theorem. Let $(X,E,\mathcal{D},P)$ be of the bridge type, and let $p=(x,y,z)$ be an I'-prepared system of parameters verifying the assumptions of (2.5.2). Let $G|_{s+1}$ be a partial realization of the reduction game beginning at $(X,E,\mathcal{D},P)$ such that it follows the standard winning strategy with respect to p until the step s (see(4.2.6)) and such that $P(s+1) \notin$ strict transform of $z=0$ and stat(s+1) is not a victory situa_ tion. Let $G'|_{s+1}$ be the I-equivalent realization of $G|_{s+1}$ relatively to E_2 and let $p' = (x',y',z')$ be obtained from p by very good preparation. Then $G'|_{s+1}$ follows the (s+1)-retarded general winning strategy from p'.

<u>Proof</u>. For each t, $0 \le t \le s$, let p(t) (resp. p'(t)) be a regular system of parameters obtained as in (4.2.3) (resp. as in (4.4.1)) from p(0) = p (resp. from p'(0) = p'). by induction, one can prove that p'(t) is a strong (very) good preparation of p(t). Let us denote

$$\Delta(t) = \Delta(\mathcal{D}(t),E(t);p(t)) \tag{4.4.4.1}$$
$$\Delta'(t) = \Delta(\mathcal{D}(t),E'(t);p'(t))$$

where E'(t) is obtained as in the proof of (4.3.2).

If the main vertex $(\alpha(t),\beta(t))$ of $\Delta(t)$ is well prepared, then the thesis may be deduced, since T-3 must be made simultaneously by both strategies ($(\alpha(t),\beta(t))$ is the main vertex of $\Delta'(t)$ too) and if T-4 is made by the standard winning strategy of (4.2.3) then it may be made by the (s+1)-retarded general winning strategy, since in this case

$$\Delta(t) = \Delta'(t) = (\alpha(t),\beta(t)) + \mathbb{R}_o^2. \tag{4.4.4.2}$$

Assume that $(\alpha(t),\beta(t))$ is not well prepared. Since $(\alpha(0),\beta(0))=(0,1+1/r)$ following the movement of the polygons $\Delta(l)$, $0 \le l \le t$, one deduces that $(\alpha(t),\beta(t))=$ $=(0,1)$, $(1,1)$ or $(1,0)$. Since all possible vertices of ordinate $< 1+1/r$, respectively of abscissa $> 1-1/r$, have ordinate ≤ 1, respectively > 1, then one deduces by the movement of the polygon that if $(\alpha(t),\beta(t))= (0,1)$, or $(1,1)$, then $\pi(l)$ is given by $(T\text{-}1,\zeta)$ or T-3 for $l=1,\dots,t$ (always from p(l-1)). Moreover if $(\alpha(t),\beta(t)) = (0,1)$, one would have a contradiction because this implies type zero for stat(t) (remark that $e(E(t)) = 2$). If $(\alpha(t),\beta(t)) = (1,1)$ and $t < s$, then $\pi(t+1)$ is given by T-3 and a contradiction is obtained as above; if t=s, clearly the center must be the curve given by $(x(t),z(t)) = (x'(t),z'(t))$ (recall that p'(t) is a strong normalization and very good preparation of p(t)) in both strategies. If $(\alpha(t),\beta(t)) = (1,0)$ and $t < s$, then $\pi(t+1)$ is given by T-3 and the adapted order decreases, if t=s, one can reason as above.

(4.4.5) Corollary. There is a winning strategy for the reduction game beginning at the bridge type.

Proof. One can follow the standard winning strategy of (4.2.3) until a point P(s+1) $\notin$ strict transform of E_2 is obtained. Now, in view of the above theorem, one can apply (4.4.3).

(4.4.6) The above corollary ends the proof of the existence of a winning strategy for the reduction game beginning at the type one.

- V -

TYPES TWO AND THREE

0. INTRODUCTION

In view of the above chapters, it remains to study the reduction game when it begins by a type two or three. In this chapter the proof of the main result (I.(4.2.9)) will be completed by studying these types.

In the sequel "victory situation" means situation of adapted order less than r, or with the dimension of the directrix equal to zero, or of the type zero, or of the type one.

The structure of this chapter is quite similar to the chapters III and IV. If one begins with type II or III and the transition is not standard, then one can obtain a type "bridge", which will be a special type III', for which one obtains the victory.

1. STANDARD TRANSITIONS FROM THE TYPES II AND III

(1.1) A winning strategy if dim Dir $(\mathcal{D},E) = 1$

(1.1.1) Theorem. Let $(X,E,\mathcal{D},P)$ be of the type III-1-1 and let $(X',E',\mathcal{D}',P')$ be a quadratic directional blowing-up and assume that it is not a victory situation. Then $(X',E',\mathcal{D}',P')$ is of the type III-1-1 or there is a strategy for the player A in order to obtain a victory situation from $(X',E',\mathcal{D}',P')$.

Proof. Let $p = (x,y,z)$ be as in (III.(1.2.15)) and let D be as in (III.(1.2.15.1)). One has T-1,0 and $\mathcal{D}'$ is generated by

(1.1.1.1) $$D' = a'x'\partial/\partial x' + b'\partial/\partial y' + c'\partial/\partial z'$$

where $a' = a/x'^r$, $b' = b/x'^{r+1} - y'a'$, $c' = c/x'^{r+1} - z'a'$. If $\nu(b')$ and $\nu(c') \geq r+1$, then one has type III-1-1 or a victory situation. Assume that $J^r(\mathcal{D}',E') \neq 0$. Since one has type four (hence 4-0 or 4-1), up to a change in y,z, one can assume that $J^r(c) = (\underline{x})$. Let $In^r(a) = \phi(\underline{y},z)$, then $In^r(a') = \phi(\underline{y}',\underline{z}') + \underline{x}'(\ldots)$. Put $p'_1 = (x'_1,y'_1,z'_1) = (z',y',x')$. Assume that A chooses the quadratic center, then B must choose T-1, ζ or T-2 from p'_1. If T-2 one obtains $J^r(\mathcal{D}'',E'') \neq 0$, $e(E'') = 2$ and then a victory situation. If T-1,ζ, one can assume that $\zeta = 0$ without loss of generality. Then the strict transform $\mathcal{D}''$ is generated by

(1.1.1.2) $$D'' = a''_1 x''_1 \partial/\partial x''_1 + b''_1 \partial/\partial y''_1 + c''_1 z''_1 \partial/\partial z''_1$$

where $a''_1 = c'/x_1''^r$, $b''_1 = b'/x_1''^r - y_1''a_1''$, $c''_1 = a'/x_1''^{r-1} - a_1''$. If $\nu(b_1'') = r$, one can reason as above. Assume that $\nu(b_1'') \geq r+1$, then

(1.1.1.3)
$$In^r(a_1'') = \underline{z}_1''^r + x_1''(\ldots)$$
$$In^r(c_1'') = \lambda x_1'' y_1''^{r-1} - z_1''^r + x_1''(\ldots) \qquad \lambda \neq 0$$

where $\phi(y,x) = y^r + \lambda xy^{r-1}$. From (1.1.1.3) one deduces that $\dim \mathrm{Dir}(\mathcal{D}'',E'') = 0$.

(1.1.2) Theorem. Let $(X',E',\mathcal{D}',P')$ be a quadratic directional blowing up of $(X,E,\mathcal{D},P)$ and assume that it is not a victory situation. Then:

a) If $(X,E,\mathcal{D},P)$ is of the type II-1-1, then $(X',E',\mathcal{D}',P')$ is of the type III-1-1 or there is a strategy for the player A in order to obtain the victory from $(X',E',\mathcal{D}',P')$.

b) If $(X,E,\mathcal{D},P)$ is of the type II-1-2, then $(X',E',\mathcal{D}',P')$ is of the type

II-1-2 or II-1-1.

c) If $(X,E,\mathcal{D},P)$ is of the type III-1-2, then $(X',E',\mathcal{D}',P')$ is of the type II-1-2 or II-1-1.

Proof. a) Let $p = (x,y,z)$ be as in (III.(1.2.10)) and let D be as in (III.(1.2.10.1)). Assume that $In(a) = \phi(\underline{y}+\zeta\underline{x},\underline{z})$ and $In(b) = \psi(\underline{y}+\zeta\underline{x},\underline{z})$. One has T-1, ζ and $\mathcal{D}'$ is generated by

(1.1.2.1) $$D' = a'x'\partial/\partial x' + b'\partial/\partial y' + c'\partial/\partial z'$$

where $a' = a/x'^r$, $b' = (b-a)(y'-\zeta)/x'^r$, $c' = c/x'^{r+1}-z'a'$. If $\nu(b')$ and $\nu(c')\geq r+1$, then $\phi=\psi$ and one has type III-1-1 or a victory situation (remark that $In^r(a') = \phi(\underline{y}',\underline{z}')+\underline{x}'(\ldots))$. Assume that $J^r(\mathcal{D}',E') \neq 0$. Since one has type four, then $\underline{x} \in J^r(\mathcal{D},E)$. Assume first $\phi-\psi \neq 0$. Then

(1.1.2.2) $$In^r(b') = (\phi-\psi)(\underline{y}',\underline{z}') + \underline{x}'(\ldots)$$

and necessarily $(\phi-\psi)(\underline{y}',\underline{z}') \neq \mu\,\underline{y}'^r$ since otherwise one has type zero. Then, up to a change $z'_1 = z' + \lambda y'$, one can assume that $J^r(c') = (\underline{x})$. Now, reasonning as in (1.1.1), by T-2 one obtains a victory situation and by $(T-1,\zeta)$ one obtains $\nu(b_1'') = r$, hence a victory situation. Assume now $\phi-\psi = 0$, then ψ is not a power of a linear form and one can reason as in (1.1.1)

b) Let $p = (x,y,z)$ be as in (III.(1.2.10)). One has (T-1,0). If $J^r(\mathcal{D}',E') = 0$, then one has type II-1-1, or II-1-2 (i.e., the transition is standard, see (III.(2.4.2)) or a victory situation. If $J^r(\mathcal{D}',E') \neq 0$, then one has a victory situation since $e(E') = 2$.

c) One can reason as in b).

(1.1.3) Corollary. There is a winning strategy for the reduction game beginning at $(X,E,\mathcal{D},P)$ of the type II or III with dim Dir $(\mathcal{D},E) = 1$.

Proof. One can see that in each of the above cases, after making a blowing-up centered in a permissible curve tangent to the directrix, then the player A wins. Assume that there is no such a curve in each step of a realization of the game. If the realization is not finite, then, in view of (1.1.1) and (1.1.2) it is stabilished in an infinite sequence of transitions II-1-2 $\mapsto$ II-1-2 or III-1-1 $\longmapsto$ III-1-1, then the permissible curve must to exist, since this transitions correspond to the infinitely near points of some regular curve (see(I.(3.3)).

(1.2) Invariants for the standard transitions

(1.2.1) Remark. Let $(X,E,\mathcal{D},P)$ be of the type II-2 and let $p = (x,y,z)$ be a normalized base. Then $\mathcal{D}$ is generated by

(1.2.1.1) $$D = ax\partial/\partial x + b\partial/\partial y + c\partial/\partial z$$

and

(1.2.1.2) $$\mathrm{Exp}\ (D,p) = \mathrm{Exp}\ (a) \cup \mathrm{Exp}\ (b/y) \cup \mathrm{Exp}\ (c/z)$$

compare with (II.(2.2.4.2)) for the type zero and $e(E) = 1$. One can deduce results like in (II.(3.3)) and good preparation may be defined as in (II.(3.3.7)).

(1.2.2) Definition. Let $(X,E,\mathcal{D},P)$ be of the type III-2 and let $p=(x,y,z)$ be a well prepared base. The base p is "strongly well prepared" iff

a) For each vertex $(\alpha,\beta) \in \Delta_+(\mathcal{D},E,p)$, such that $(\alpha,\beta) \in \mathbb{Z}_0^2$ and $\beta \geq 1$, there is no change $z_1=z+\lambda x^\alpha y^\beta$ which may dissolve this vertex of $\Delta_+(\mathcal{D},E,p)$

b) For each vertex $(n,0) \in \Delta_+(\mathcal{D},E,p)$ there is no change $z_1=z+\lambda x^n$ such that $\Delta(\mathcal{D},E,p)$ is not modified and $(n,0)$ disappears in $\Delta_+(\mathcal{D},E,p)$.

(1.2.3) Lemma. a) If p is well prepared, then one can obtain a base p' strongly well prepared by a sequence of changes $z_1=z+\lambda x^\alpha y^\beta$, such that $\Delta(\mathcal{D},E,p)= = \Delta(\mathcal{D},E,p')$.

b) If p is strongly well prepared, then $J(\mathcal{D},E) = (\underline{z}+\lambda\underline{x})$.

Proof. a) It follows from the results in (II.(3.3)).

b) If $(0,1) \in \Delta_+(\mathcal{D},E,p)$, since $\dim \mathrm{Dir}(\mathcal{D},E) = 2$, then a change $z_1 = = z + \lambda y$ dissolves this vertex.

(1.2.4) Remark. Very good preparation is defined as in (II.(3.4.5)) by putting "strongly good preparation" instead of "good preparation". Also one can obtain a strongly very well prepared base from a strongly well prepared base in the usual way (see II.(3.4.6)).

(1.2.5) Theorem. Let $(X,E,\mathcal{D},P)$ be of the type II-2 or III-2 and let $p = (x,y,z)$ be a strongly (very if type III) well prepared base. Assume that $(X',E',\mathcal{D}',P')$ is a directional blowing-up such that the center follows the 0-retarded standard winning strategy (see III.(2.8.4)). Then one of the following possibilities is satisfied:

a) The transition is not standard.

b) $\dim \operatorname{Dir}(\mathcal{D}',E') = 1$.

c) The transition is standard and there is a strongly (very if type III) well prepared base $p' = (x',y',z')$ such that the invariant $(\beta',e(E'),\varepsilon',\alpha')$ is strictly smaller than $(\beta,e(E),\varepsilon,\alpha)$ where $\beta = \beta(\Delta(\mathcal{D},E,p))$, $\beta' = \beta(\Delta(\mathcal{D}',E',p'))$ etc.

Proof. Assume that a) and b) are not satisfied. Assume first that $(X,E,\mathcal{D},P)$ is of the type II-2. Then $\underline{z} = J(\mathcal{D},E)$ and the transformation is given by (T-1,0), $(T\text{-}1,\zeta)$, $\zeta \neq 0$, T-2, T-3 or T-4. If one has not $(T\text{-}1,\zeta)$, $\zeta \neq 0$, then the result follows from (III.(2.6.2)) and (III.(2.2.5)). If one has $(T\text{-}1,\zeta)$, $\zeta \neq 0$, one deduces (see II.(4.4.3)) that if $\beta' \geq \beta$, then the main vertex of $\Delta(\mathcal{D}',E',p'_1)$, where p'_1 is obtained from p by $T\text{-}1,\zeta$, is $(\alpha+\beta-1,\beta)$ and it is well prepared, then it is preserved after strong very good preparation (see II.(3.4.3)) vgr.).

Assume now that $(X,E,\mathcal{D},P)$ is of the type III-2. Assume first that $\underline{z} = J(\mathcal{D},E)$. Then one can reason as in (II.(4.3.3)). Let us assume that $J(\mathcal{D},E) = (\underline{z}+\lambda\underline{x})$, where $\lambda \neq 0$. Since p is well prepared, one can see that a change $z_1 = z + \lambda x$ does not modify the main vertex of $\Delta(\mathcal{D},E,p)$ nor the fact that it is well prepared (i.e. p is prepared until the first vertex) unless (1,0) is the main vertex, but in this last case, one cannot have a standard transition: then the result follows if one has T-2, T-4 (T-3 is not possible). Assume now that one has

$(T-1,\zeta)$, and that $(1,0)$ is not the main vertex of $\Delta(\mathcal{D},E,p)$, then $\varepsilon \leq 1$, and after $y_1 = y+\zeta x$, $z_1 = z+\lambda x$ this property holds and one obtains $\beta' < \beta$.

(1.2.6) Corollary. Let G be a realization of the reduction game beginning at $(X,E,\mathcal{D},P)$ of the type II or III and let us assume that all the transitions in G are standard (unless the last one) and that for each s, $G|_s$ follows the l-retarded standard winning strategy (definition as in (III.(2.8.4))). Then G is finite.

Proof. One can reason as in (III.(2.8.5)).

2. NO STANDARD TRANSITIONS FROM II AND III

(2.1) Another invariant of transversality

(2.1.1) An invariant of transversality for the types II' and III' will be introduced in order to be able for defining the bridge type mentioned in the introduction. This invariant is also useful for the study of the type II' and III'.

(2.1.2) Definition. Let $(X,E,\mathcal{D},P)$ be of the type II' (resp. III'). The invariant $\tau=\tau(X,E,\mathcal{D},P)$ is defined to be zero iff there is a decomposition $E = E_1 \cup E_2$, E_i normal crossings divisor, $i = 1,2$, $e(E_2) = 1$ such that there is $\phi \in R$, with $I(E_1) \subset \phi .R$ in such a way that

(2.1.2.1) $$J^r(D(\phi)/\phi) \not\subset J(E_1)$$

Otherwise $\tau = 1$.

(2.1.3) Remark. Assume that $(X,E,\mathcal{D},P)$ is of the type II' and $p = (x,y,z)$ is a r.s. of p. suited for (E,P) such that $\mathcal{D}$ is generated by

(2.1.3.1) $$D = ax\partial/\partial x + by\partial/\partial y + cz\partial/\partial z$$

Then $\tau = 0$ iff p may be chosen in such a way that $J^r(a) + J^r(b) \not\subset (\underline{x},\underline{y})$. If $\tau = 1$, then $J^r(a) \neq 0$ (resp. $J^r(b), J^r(c)$) implies $J^r(a) = (\underline{x})$, (resp. $J^r(b) = (\underline{y})$, $J^r(c)$ =

$= (\underline{z}))$.

Assume that $(X,E,\mathcal{D},P)$ is of the type III' and $p = (x,y,z)$ is such that $I(E) = (xz)$, and $\mathcal{D}$ is generated by

$$D = ax\partial/\partial x + b\partial/\partial y + cz\partial/\partial z \tag{2.1.3.2}$$

Then $\tau = 0$ iff p may be chosen in such a way that $J^r(a) \not\subset (\underline{x})$. If $\tau = 1$, then $J^r(a) \neq \neq 0$ (resp. $J^r(c)$), implies $J^r(a) = (\underline{x})$, resp. $J^r(c) = (\underline{z})$.

(2.1.4) Definition. Let $(X,E,\mathcal{D},P)$ be of the type II' or III', a regular system of parameters $p = (x,y,z)$ is "normalized" iff p is suited for (E,P), $I(E) \subset (x,z)$ and and if $\tau = 0$ one has the property of (2.1.3) and if $\tau = 1$ one has that $J^r(c) \neq 0$ (hence $J^r(c) = (\underline{z})$).

(2.1.5) Lemma. With notations as in (2.1.2), if $\tau = 0$, then $(X,E_1,\mathcal{D},P)$ is of the type II or III (if $(X,E,\mathcal{D},P)$ is respectively II' or III') and $r = \nu(\mathcal{D},E_1,P)$.

Proof. See (2.1.3)

(2.1.6) Definition. Let $(X,E,\mathcal{D},P)$ be of the type III'. It is of the type III-bridge iff $\tau = 0$ and there is a normalized system of parameters $p = (x,y,z)$ such that if $E = E_1 \cup E_2$ is the decomposition of (2.1.2), then one has that

a) $\dim \mathrm{Dir}\,(\mathcal{D},E_1) = 2$.

b) $(h,i,j) \in \mathrm{Exp}\,(D,E_1,p)$, $j < r \Rightarrow h \geq 1$.

c) $(1,r,0) \in \mathrm{Exp}\,(D,E_1,p)$.

(2.2) Quadratic no standard transitions from III

(2.2.1) Assume that $(X,E,\mathcal{D},P)$ is of the type III-2 and that $(X',E',\mathcal{D}',P')$ is a quadratic blowing-up of $(X,E,\mathcal{D},P)$ such that it is not a victory situation nor corresponds to a standard transition. Let $p = (x,y,z)$ be a base verifying the properties of the lemma (III.(1.2.15)). Then the transformation is given by $(T\text{-}1,\zeta)$ or by T-2.

If it is given by T-2, one has $e(E')=2$ and if $J^r(\mathcal{D}',E')=0$ one has victory situation or a standard transition and if $J^r(\mathcal{D}',E')\neq 0$ one has a victory situation. Then one can assume that the transformation is given by $(T\text{-}1,\zeta)$ and, without loss of generality, by T-1,0. Then, if $\mathcal{D}$ is generated by $D=ax\partial/\partial x+b\partial/\partial y+c\partial/\partial z$, then $\mathcal{D}'$ is generated by $D'=a'x'\partial/\partial x'+b'\partial/\partial y'+c'\partial/\partial z'$, where

(2.2.1.1) $$a'=a/x'^r;\ b'=b/x'^{r+1}-y'a';\ c'=c/x'^{r+1}-z'a'$$

This implies that

(2.2.1.2) $$a'=z'^r+x'(\dots)$$
$$b'=\Phi(y',z')+x'(\dots)-y'a'$$
$$c'=\Psi(y',z')+x'(\dots)-z'a'$$

(2.2.2) Assume now that $J_H(\mathcal{D}',E')\neq 0$. Since one has not a victory situation, then $J_H(\mathcal{D}',E')=(\underline{x}')$. This implies that the following (quadratic) transformation, must be given by T-2 (see (2.2.1.2)). Then $\mathcal{D}''$ is generated by $D''=a''x''\partial/\partial x''+b''y''\partial/\partial y''+c''\partial/\partial z''$, where

(2.2.2.1) $$a''=a'/y''^{r-1}-b'';\ b''=b'/y''^r;\ c''=c'/y''^r-z''b''.$$

If $\nu(c'')=r$, one obtains a victory situation. Then $\nu(c'')\geq r+1$, hence $\nu(c')\geq r+1$. But this implies that $\underline{y}+\lambda\underline{x}+\mu\underline{z}\in J^r(\mathcal{D}',E')=J^r(b')$ and one has type zero (or better). Then one can assume $J_H(\mathcal{D}',E')\neq 0$. This implies $J^r(\mathcal{D}',E')=(\underline{x})$. In particular, this implies that Φ and Ψ are homogeneous of degree r+1.

(2.2.3) Put $p_1=(x_1,y_1,z_1)=(y'',z'',x'')$. From (2.2.2.1) and (2.2.1.2) one has that

(2.2.3.1) $$a_1=b''=\lambda z_1^r+x_1(\Phi(1,y_1)-y_1^r)+x_1z_1(\dots)$$
$$b_1=c''=-\lambda y_1z_1^r+x_1y_1((\Psi(1,y_1)/y_1)-2y_1^r-\Phi(1,y_1))+x_1z_1(\dots)$$
$$c_1=a''=-\lambda z_1^r-x_1\Phi(1,y_1)+x_1z_1(\dots)$$

where $\lambda\neq 0$. Assume that $\Phi(1,y)=\mu y^{r-1}+\gamma y^r+y^{r+1}$. One can assume that $\mu=0$, since otherwise the dimension of the directrix is zero. Since γ and $\gamma-1$ are not simultaneously zero, from (2.2.3.1) one can deduce that and if $J^r(a_1)=(\underline{z}_1)$, then one has a type III-bridge.

(2.2.4) Proposition. With notations as above, there is a bijection ϕ : Exp (D,E,p) $\longrightarrow$ Exp (D'',E''_1,p_1) given by

(2.2.4.1) $$\phi(h,i,j) = (h+2(i+j-r)+1, j, h+i+j-r)$$

where $I(E''_1) = (x_1)$.

Proof. It follows from (2.2.1.1) and (2.2.2.1).

(2.3) No standard transitions from III and II

(2.3.1) Assume that $(X,E,\mathcal{D},P)$ is of the type III-2 or II-2 and that π: $X' \to X$ induces a directional blowing-up with permissible center tangent to the directrix that the transition is not standard and $(X',E',\mathcal{D}',P')$ is not a victory situation. Then $J^r(\mathcal{D}',E') \neq 0$ and $e(E') = 1$. In view of the above paragraph, one has one of the following possibilities.

a) π quadratic and $(X,E,\mathcal{D},P)$ is of the type II-2.

b) π monoidal.

(2.3.2) Assume that π is quadratic. Let $p = (x,y,z)$ be a regular system of parameters as in (III.(1.2.8)). Since $e(E')$, π must be given by $(T-1,\zeta)$, $\zeta \neq 0$. Assume that $\mathcal{D}$ is generated by $D = ax\partial/\partial x + by\partial/\partial y + c\,\partial/\partial z$. There are two possibilites: $J^r(a) = (\underline{z})$ or $J^r(a) = 0$. If $J^r(a) = \underline{z}$, then after making $y_1 = y + \zeta x$, one can reason as in the precedent paragraph and one obtains a bridge type in the following quadratic blowing-up.

(2.3.2) Assume that π is monoidal. Without loss of generality, one can choose $p = (x,y,z)$ as in (III.(1.2.8)) or as in (III.(1.2.15)), depending on type II or III, with the additionnal property that the center of the blowing-up is given by (x,z) or by (y,z). Then the transformation must be given by T-3 or T-4 from p, since

$J^r(\mathcal{D},E) = (\underline{z})$. Moreover, since one must to have $e(E') = 1$ (otherwise one has a victory situation), necessarily $(X,E,\mathcal{D},P)$ is of the type III-2 and π is given by T-3. Assume that $\mathcal{D}$ is generated by $D = ax\partial/\partial x + b\partial/\partial y + c\partial/\partial z$, where $J^r(a) = (\underline{z})$, after T-3, $\mathcal{D}'$ is generated by $D' = a'x'\partial/\partial x' + b'\partial/\partial y' + c'\partial/\partial z'$ where

(2.3.2.1) $$a' = a/x'^r;\ b' = b/x'^r,\ c' = c/x'^{r+1} - z'a'$$

Now, one can reason as in (2.2.2) in order to prove that $J_H(\mathcal{D}',E') = 0$. Then $J^r(\mathcal{D}',E') = (\underline{x}')$ and $J^r(b') = (\underline{x}')$ and $\nu(c') \geq r+1$ (see (2.2.2)). Remark that

(2.3.2.2) $$a' = z'^r + x'(\ldots) + z'^r.y'(\ldots)$$
$$b' = \lambda x'^r + x'\phi(x',y',z') + z'^r(\ldots) \qquad \lambda \neq 0$$

where $\nu(\phi) \geq r$. Reasonning as in (2.2.2), one has to make T-2 (if A chooses the quadratic center). Let $p'' = (x'',y'',z'')$ be the obtained base and put $p_1 = (x_1,y_1,z_1) = (y'',z'',x'')$ as in (2.2.3). Then $\mathcal{D}''$ is generated by

(2.3.2.3) $$D'' = a_1x_1\partial/\partial x_1 + b_1\partial/\partial y_1 + c_1z_1\partial/\partial z_1$$

where

$$a_1 = b'' = b'/y''^r = \lambda z_1^r + x_1z_1(\ldots) + x_1^2y_1^r(\ldots)$$
$$b_1 = c'' = c'/y''^r - z''b'' = -\lambda y_1z_1 + x_1(\ldots)$$
$$c_1 = a'' = a'/y''^{r-1} - b'' = -\lambda z_1^r + x_1y_1^r + x_1z_1(\ldots) + x_1^2y_1^r(\ldots)$$

Then, one has a type III-bridge (if $\nu(b_1) \geq r+1$ and dim Dir $(\mathcal{D}'',E'') = 2$). If dim Dir $(\mathcal{D}'',E'') = 1$ one has a type III' with $\tau = 0$, and after the following transformation (always T-2) then one obtains a victory situation (the order drops).

The following theorem resumes the computations made up till now.

(2.3.3) <u>Theorem</u>. If the reduction game begins at the type II or III, then, there is a strategy for the player A in order to win or to obtain a type III-bridge.

(2.4) A winning strategy for the type III-bridge

(2.4.1) <u>Theorem</u>. Assume that the reduction game begins at $(X,E,\mathcal{D},P)$ which is of the type III-2 and such that there is a normalized system of parameters $p = (x,y,z)$ verifying that

a) $\forall (h,i,j) \in \mathrm{Exp}(D,E,p)$, $j < r \Rightarrow h \geq 1$. Moreover, if $j=-1$ and $i \geq r+1$, then $h \geq 2$.

b) $(1,r,0) \in \mathrm{Exp}(D,E,p)$.

Then there is a winning strategy for the reduction game beginning at $(X,E,\mathcal{D},P)$.

Proof. Assume first that the player A chooses the quadratic center. One can modify p is such a way that $J(\mathcal{D},E) = (\underline{z})$, without touching the conditions a) and b). Then the transformation may be given by T-2 or T-1,ζ. By making $y_1 = y + \zeta x$ (which does not modify a) and b)) one can assume that the transformation is given by T-2 or T-1,0.

If one has T-2, since $(1/r,1) \in \Delta(\mathcal{D},E,p)$, one deduces that the transition is not standard and since $e(E') = 2$, then the player A wins.

Assume that the transformation is given by T-1,0. First, if the transition is not standard, one can apply the results of (2.2) and if the player A has not won, then by (2.2.4), for same base p_1 one has that

$$(1,r,0) = (1,0,1) \in \mathrm{Exp}(D'',E''_1,p_1) \tag{2.4.1.1}$$

and if $r \geq 3$, the player A has won. If $r = 2$, looking at (2.3.1), from (2.4.1.1) one deduces that after the following quadratic transformation (necessarily T-2), the dimension of the directrix becomes zero. If the transition is standard, then the strict transform $(X',E',\mathcal{D}',P')$ satisfies once more the properties a) and b). One can repeat.

If the processus does not stop, after a change $y_1 = y + \sum \zeta_i x^i$, $z_1 = z + \sum \mu_i x^i$ which does not modify a), b), one can assume that one has an infinite sequence of transformations T-1,0. This implies that $(1/r,1)$ is the only vertex of $\Delta(\mathcal{D},E,p)$. Then the player A wins by choosing the center given by (y,z), since the transformation is not standard and $e(E') = 2$.

(2.4.2) Theorem. There is a winning strategy for the reduction game beginning at a type III-bridge.

Proof. Let $(X,E,\mathcal{D},P)$ be of the type III-bridge and let $E = E_1 \cup E_2$ and $p = (x,y,z)$ be as in (2.1.6). Assume that the player A chooses the quadratic transformation, then $(X',E',\mathcal{D}',P')$ is a victory situation, a bridge type or it verifies a) and b) of (2.4.1). Now one can reason in the same way as in (2.4.1).

(2.4.3) Corollary. There is a winning strategy for the reduction game beginning at a type II or III.

3. TYPES II' AND III'

(3.1) The case $\tau = 0$

(3.1.1) In this paragraph, types II, III and the type III-bridge will be considered as "victory situations" as well as the "victory situations" of the introduction.

(3.1.2) Theorem. Let $(X,E,\mathcal{D},P)$ be of the type II' or III' with $\tau = 0$ and let $p = (x,y,z)$ be a normalized system of parameters. Let $(X',E',\mathcal{D}',P')$ be a directional blowing-up with a permissible center tangent to the directrix. Then one of the following possibilities is satisfied

a) There is a winning strategy for the reduction game beginning at $(X',E',\mathcal{D}',P')$ or it is a victory situation.

b) If the transformation is quadratic, then it is given by $(T\text{-}1,\zeta)$ or T-2 from p, $(X',E',\mathcal{D}',P')$ is of the types II' or III', $\tau = 0$ and the obtained base is normalized.

c) If the transformation is monoidal, then (up to a change $y_1 = y + \sum \zeta_i x^i$ in the case III') it is given by T-3 or T-4, $(X',E',\mathcal{D}',P')$ is of the types II' or III' $\tau = 0$, and the obtained base is normalized.

Proof. Assume that $(X',E',\mathcal{D}',P')$ is not a victory situation. First, let us assume that the transformation is quadratic. There are two possibilities: $P' \in$ stric transform of $\underline{z} = 0$ or not. Assume that $P' \notin$ strict transform of $\underline{z}=0$. Then, $(X',E',\mathcal{D}',P')$ may be considered as the strict transform of $(X,E_1,\mathcal{D},P)$, where $I(E_1)= =(x)$ or $I(E_1)=(xy)$, depending on $e(E) = 2$ or $e(E) = 3$. Now, in view of (2.1.5), the result a) follows from the computations of the above sections.

Assume that $P \in$ strict transform of $\underline{z} = 0$. Then the transformation is given by $(T-1, \zeta)$ or T-2 from p. If $\mathcal{D}$ is generated by $D = ax\partial/\partial x + b\partial_y + cz\partial/\partial z$ where $\partial_y = \partial/\partial y$ or $y\partial/\partial y$ depending on $e(E) = 2$ or 3. Assume that $\mathcal{D}'$ is generated by $D'= = a'x'\partial/\partial x' + b'\partial_{y'} + c'z'\partial/\partial z'$, necessarily $J^r(\mathcal{D}',E') = 0$, otherwise one has a). From the equations of the transformation, one can deduce that

(3.1.2.1) $$J^r(a) \text{ (resp. } J^r(a)+J^r(b)) \not\subset (\underline{x}) \text{ (resp. } (\underline{x},\underline{y}) \Rightarrow$$
$$\Rightarrow J^r(a') \text{ (resp. } J^r(a')+J^r(b')) \not\subset (\underline{x}') \text{ (resp. } (\underline{x}',\underline{y}'))$$

And if one has not a type II or better (instead of III') in the case $e(E') = 2$, one deduces that $\tau = 0$, $(X',E',\mathcal{D}',P')$ is of the type II' or III' and $p' = (x',y',z')$ is normalized.

Now let Y be a permissible curve tangent to the directrix, since $\tau = 0$ and p is normalized, necessarily $I(Y) \neq (x,y)$ and $I(Y) = (x,z)$ or $I(Y) = (y_1,z)$ where y_1 is as in c). Now, one can can reason as above.

(3.1.3) <u>Corollary</u>. There is a winning strategy for the reduction game beginning at a type II' or III' with $\tau = 0$.

<u>Proof</u>. The above theorem allows us to use the techniques of the polygons in the usual way: for the case III' one can prepare the polygon by means of changes $y_1 = y+\zeta x^n$. The strategy is as usual: to choose (x,z) if it is permissible, otherwise (y,z) if it is permissible, otherwise the quadratic center.

(3.2) <u>The case $\tau = 1$</u>

(2.3.1) <u>Definition</u>. Let $(X,E,\mathcal{D},P)$ be of the type II' or III' with $\tau = 1$ and let $p = (x,y,z)$ be a normalized system of parameters p is "strongly normalized" iff $J^r(D(z)/z) \neq 0$ (hence $J^r(D(z)/z) = (\underline{z})$) where D generates $\mathcal{D}$

(3.2.2) Theorem. Let $(X,E,\mathcal{D},P)$ be of the type II' or III' with $\tau = 1$ and let $p = (x,y,z)$ be a strongly normalized system of parameters. Let $(X',E',\mathcal{D}',P')$ be a directional blowing-up with a permissible center tangent to the directrix. Then one of the following possibilities is satisfied.

a) There is a winning strategy for the reduction game beginning at $(X',E',\mathcal{D}',P')$ or it is a victory situation.

b) $(X',E',\mathcal{D}',P')$ is of the type II' or III' with $\tau = 1$, the transformation is given by $(T\text{-}1,\zeta)$, T-2, T-3 or T-4 (up to a change $y_1 = y + \sum \zeta_i x^i$ in the last case) from p and the obtained base $p' = (x',y',z')$ is strongly normalized.

Proof. One has that $J(\mathcal{D},E) \ni \underline{z}$ and one can deduce that the transformation must be given by $(T\text{-}1,\zeta)$, T-2, T-3 or T-4 if a) is not satisfied. Looking at the equations, since $e(E') \geq 2$ one must to have types II' or III' with $\tau = 1$ (otherwise a)). Moreover, p' is normalized. If $J^r(D'(z')/z') = 0$ this implies that $J^r(D(x)/x) \not\subset (\underline{x})$ or that $J^r(D(x)/x) + J^r(D(y)/y) \not\subset (\underline{x},\underline{y})$ in the cases III' and II' respectively, then p' is strongly normalized.

(3.2.3) Corollary. There is a winning strategy for the reduction game beginning at the types II' or III' with $\tau = 1$.

Proof. As in (3.1.3) one can take the usual strategy over the polygon. The control is assured until one obtains a victory situation.

(3.2.4) The above corollary ends the proof of the main result (I.(4.2.9)).

-REFERENCES-

|1| ABHYANKAR, S.S.. "Desingularization of plane curves". Proc. of Symp. in Pure Math., A.M.S. vol. 40. Arcata 1981.

|2| CANO, F.. "Teoría de distribuciones sobre variedades algebraicas". Mem. y Mon. Inst. Jorge Juan, C.S.I.C., Madrid. 1983.

|3| ________. "Desingularization of plane vector fields". Transac. of the A.M.S. Vol. 296, 1, pp. 83/93.

|4| ________. "Techniques pour la désingularisation des champs de vecteurs". Proc. la Rábida (1984). Huelva. To appear in "Travaux en cours". Hermann.

|5| ________. "Jeux de résolution pour les champs de vecteurs en dimension trois". Publ. Ecole Polytechnique. Palaiseau. 1985.

|6| CERVEAU, D. - MATTEI, J.F.. "Formes holomorphes intégrables singulières". Asterisque 97. 1982.

|7| COSSART, V.. "Forme normale d'une fonction en dimension trois". Proc. la Rábida (1984). Huelva. To appear in "Travaux en cours". Hermann.

|8| GIRAUD, J.. "Forme normale d'une fonction sur une surface de caractéristique positive". Bull. Soc. Math. France, 111, 1983, p. 109-124.

|9| ________. " Condition de Jung pour les revêtements radiciels de hauteur un" Proc. Algebraic Geometry, Tokyo/Kyoto 1982. Lecture Notes in Math. nº 1016. Springer Verlag. 1983. p. 313-333.

|10| HIRONAKA, H.. "Desingularization of excellent surfaces". Adv. Sci. Sem. in Alg. Geom. bowdoin College (1967). Appeared in Lecture Notes in Math. nº1101 Springer Verlag (1984).

|11| ________. "Resolution of the singularities of an algebraic variety over a field of characteristic zero". Ann. of Math. 79, 109/326. 1964.

|12| SANCHEZ-GIRALDA, T.. "Caractérisation des variétés permises d'une hypersurface algebroïde". C.R. Ac. Sci. Paris L. 285. 1977.

|13| SEIDENBERG, A.." Reduction of the singularities of the differential equation Ady=Bdx". Am. J. of Math. 1968, p. 248/269.

-INDEX-

... ...

LECTURE NOTES IN MATHEMATICS
Edited by A. Dold and B. Eckmann

Some general remarks on the publication of monographs and seminars

In what follows all references to monographs, are applicable also to multiauthorship volumes such as seminar notes.

1. Lecture Notes aim to report new developments - quickly, informally, and at a high level. Monograph manuscripts should be reasonably self-contained and rounded off. Thus they may, and often will, present not only results of the author but also related work by other people. Furthermore, the manuscripts should provide sufficient motivation, examples and applications. This clearly distinguishes Lecture Notes manuscripts from journal articles which normally are very concise. Articles intended for a journal but too long to be accepted by most journals, usually do not have this "lecture notes" character. For similar reasons it is unusual for Ph.D. theses to be accepted for the Lecture Notes series.

   Experience has shown that English language manuscripts achieve a much wider distribution.

2. Manuscripts or plans for Lecture Notes volumes should be submitted either to one of the series editors or to Springer-Verlag, Heidelberg. These proposals are then refereed. A final decision concerning publication can only be made on the basis of the complete manuscripts, but a preliminary decision can usually be based on partial information: a fairly detailed outline describing the planned contents of each chapter, and an indication of the estimated length, a bibliography, and one or two sample chapters - or a first draft of the manuscript. The editors will try to make the preliminary decision as definite as they can on the basis of the available information.

3. Lecture Notes are printed by photo-offset from typed copy delivered in camera-ready form by the authors. Springer-Verlag provides technical instructions for the preparation of manuscripts, and will also, on request, supply special staionery on which the prescribed typing area is outlined. Careful preparation of the manuscripts will help keep production time short and ensure satisfactory appearance of the finished book. Running titles are not required; if however they are considered necessary, they should be uniform in appearance. We generally advise authors not to start having their final manuscripts specially tpyed beforehand. For professionally typed manuscripts, prepared on the special stationery according to our instructions, Springer-Verlag will, if necessary, contribute towards the typing costs at a fixed rate.

   The actual production of a Lecture Notes volume takes 6-8 weeks.

.../...

4. Final manuscripts should contain at least 100 pages of mathematical text and should include

   - a table of contents
   - an informative introduction, perhaps with some historical remarks. It should be accessible to a reader not particularly familiar with the topic treated.
   - subject index; this is almost always genuinely helpful for the reader.

5. Authors receive a total of 50 free copies of their volume, but no royalties. They are entitled to purchase further copies of their book for their personal use at a discount of 33 1/3 %, other Springer mathematics books at a discount of 20 % directly from Springer-Verlag.

   Commitment to publish is made by letter of intent rather than by signing a formal contract. Springer-Verlag secures the copyright for each volume.

LECTURE NOTES

ESSENTIALS FOR THE PREPARATION OF CAMERA-READY MANUSCRIPTS

Springer
Springer-Verlag
Berlin Heidelberg New York
London Paris Tokyo

The preparation of manuscripts which are to be reproduced by photo-offset requires special care. Manuscripts which are submitted in technically unsuitable form will be returned to the author for retyping. There is normally no possibility of carrying out further corrections after a manuscript is given to production. Hence it is crucial that the following instructions be adhered to closely. If in doubt, please send us 1 - 2 sample pages for examination.

Typing area. On request, Springer-Verlag will supply special paper with the typing area outlined.

The CORRECT TYPING AREA is 18 x 26 1/2 cm (7,5 x 11 inches).

Make sure the TYPING AREA IS COMPLETELY FILLED. Set the margins so that they precisely match the outline and type right from the top to the bottom line. (Note that the page-number will lie outside this area). Lines of text should not end more than three spaces inside or outside the right margin (see example on page 4).

Type on one side of the paper only.

Type. Use an electric typewriter if at all possible. CLEAN THE TYPE before use and always use a BLACK ribbon (a carbon ribbon is best).

Choose a type size large enough to stand reduction to 75%.

Word Processors. Authors using word-processing or computer-typesetting facilities should follow these instructions with obvious modifications. Please note with respect to your printout that
i) the characters should be sharp and sufficiently black;
ii) if the size of your characters is significantly larger or smaller than normal typescript characters, you should adapt the length and breadth of the text area proportionally keeping the proportions 1:0.68.
iii) it is not necessary to use Springer's special typing paper. Any white paper of reasonable quality is acceptable.
IF IN DOUBT, PLEASE SEND US 1-2 SAMPLE PAGES FOR EXAMINATION. We will be glad to give advice.

Spacing and Headings (Monographs). Use ONE-AND-A-HALF line spacing in the text. Please leave sufficient space for the title to stand out clearly and do NOT use a new page for the beginning of subdivisions of chapters. Leave THREE LINES blank above and TWO below headings of such subdivisions.

Spacing and Headings (Proceedings). Use ONE-AND-A-HALF line spacing in the text. Start each paper on a NEW PAGE and leave sufficient space for the title to stand out clearly. However, do NOT use a new page for the beginning of subdivisions of a paper. Leave THREE LINES blank above and TWO below headings of such subdivisions. Make sure headings of equal importance are in the same form.

The first page of each contribution should be prepared in the same way. Therefore, we recommend that the editor prepares a sample page and passes it on to the authors together with these ESSENTIALS. Please take

.../...

the following as an example.

MATHEMATICAL STRUCTURE IN QUANTUM FIELD THEORY

John E. Robert
Fachbereich Physik, Universität Osnabrück
Postfach 44 69, D-4500 Osnabrück

Please leave THREE LINES blank below heading and address of the author. THEN START THE ACTUAL TEXT OF YOUR CONTRIBUTION.

Footnotes. These should be avoided. If they cannot be avoided, place them at the foot of the page, separated from the text by a line 4 cm long, and type them in SINGLE LINE SPACING to finish exactly on the outline.

Symbols. Anything which cannot be typed may be entered by hand in BLACK AND ONLY BLACK ink. (A fine-tipped rapidograph is suitable for this purpose; a good black ball-point will do, but a pencil will not). Do not draw straight lines by hand without a ruler (not even in fractions).

Equations and Computer Programs. Equations and computer programs should begin four spaces inside the left margin. Should the equations be numbered, then each number should be in brackets at the right-hand edge of the typing area.

Pagination. Number pages in the upper right-hand corner in LIGHT BLUE OR GREEN PENCIL ONLY. The final page numbers will be inserted by the printer.

There should normally be NO BLANK PAGES in the manuscript (between chapters or between contributions) unless the book is divided into Part A, Part B for example, which should then begin on a right-hand page.

It is much safer to number pages AFTER the text has been typed and corrected. Page 1 (Arabic) should be THE FIRST PAGE OF THE ACTUAL TEXT. The Roman pagination (table of contents, preface, abstract, acknowledgements, brief introductions, etc.) will be done by Springer-Verlag.

Corrections. When corrections have to be made, cut the new text to fit and PASTE it over the old. White correction fluid may also be used.

Never make corrections or insertions in the text by hand.

If the typescript has to be marked for any reason, e.g. for TEMPORARY page numbers or to mark corrections for the typist, this can be done VERY FAINTLY with BLUE or GREEN PENCIL but NO OTHER COLOR: these colors do not appear after reproduction.

Table of Contents. It is advisable to type the table of contents later, copying the titles from the text and inserting page numbers.

Literature References. These should be placed at the end of each paper or chapter, or at the end of the work, as desired. Type them with single line spacing and start each reference on a new line.
Please ensure that all references are COMPLETE and PRECISE.

Vol. 1090: Differential Geometry of Submanifolds. Proceedings, 1984. Edited by K. Kenmotsu. VI, 132 pages. 1984.

Vol. 1091: Multifunctions and Integrands. Proceedings, 1983. Edited by G. Salinetti. V, 234 pages. 1984.

Vol. 1092: Complete Intersections. Seminar, 1983. Edited by S. Greco and R. Strano. VII, 299 pages. 1984.

Vol. 1093: A. Prestel, Lectures on Formally Real Fields. XI, 125 pages. 1984.

Vol. 1094: Analyse Complexe. Proceedings, 1983. Edité par E. Amar, R. Gay et Nguyen Thanh Van. IX, 184 pages. 1984.

Vol. 1095: Stochastic Analysis and Applications. Proceedings, 1983. Edited by A. Truman and D. Williams. V, 199 pages. 1984.

Vol. 1096: Théorie du Potentiel. Proceedings, 1983. Edité par G. Mokobodzki et D. Pinchon. IX, 601 pages. 1984.

Vol. 1097: R.M. Dudley, H. Kunita, F. Ledrappier, École d'Éte de Probabilités de Saint-Flour XII – 1982. Edité par P.L. Hennequin. X, 396 pages. 1984.

Vol. 1098: Groups – Korea 1983. Proceedings. Edited by A.C. Kim and B.H. Neumann. VII, 183 pages. 1984.

Vol. 1099: C.M. Ringel, Tame Algebras and Integral Quadratic Forms. XIII, 376 pages. 1984.

Vol. 1100: V. Ivrii, Precise Spectral Asymptotics for Elliptic Operators Acting in Fiberings over Manifolds with Boundary. V, 237 pages. 1984.

Vol. 1101: V. Cossart, J. Giraud, U. Orbanz, Resolution of Surface Singularities. Seminar. VII, 132 pages. 1984.

Vol. 1102: A. Verona, Stratified Mappings – Structure and Triangulability. IX, 160 pages. 1984.

Vol. 1103: Models and Sets. Proceedings, Logic Colloquium, 1983, Part I. Edited by G.H. Müller and M.M. Richter. VIII, 484 pages. 1984.

Vol. 1104: Computation and Proof Theory. Proceedings, Logic Colloquium, 1983, Part II. Edited by M.M. Richter, E. Börger, W. Oberschelp, B. Schinzel and W. Thomas. VIII, 475 pages. 1984.

Vol. 1105: Rational Approximation and Interpolation. Proceedings, 1983. Edited by P.R. Graves-Morris, E.B. Saff and R.S. Varga. XII, 528 pages. 1984.

Vol. 1106: C.T. Chong, Techniques of Admissible Recursion Theory. IX, 214 pages. 1984.

Vol. 1107: Nonlinear Analysis and Optimization. Proceedings, 1982. Edited by C. Vinti. V, 224 pages. 1984.

Vol. 1108: Global Analysis – Studies and Applications I. Edited by Yu.G. Borisovich and Yu.E. Gliklikh. V, 301 pages. 1984.

Vol. 1109: Stochastic Aspects of Classical and Quantum Systems. Proceedings, 1983. Edited by S. Albeverio, P. Combe and M. Sirugue-Collin. IX, 227 pages. 1985.

Vol. 1110: R. Jajte, Strong Limit Theorems in Non-Commutative Probability. VI, 152 pages. 1985.

Vol. 1111: Arbeitstagung Bonn 1984. Proceedings. Edited by F. Hirzebruch, J. Schwermer and S. Suter. V, 481 pages. 1985.

Vol. 1112: Products of Conjugacy Classes in Groups. Edited by Z. Arad and M. Herzog. V, 244 pages. 1985.

Vol. 1113: P. Antosik, C. Swartz, Matrix Methods in Analysis. IV, 114 pages. 1985.

Vol. 1114: Zahlentheoretische Analysis. Seminar. Herausgegeben von E. Hlawka. V, 157 Seiten. 1985.

Vol. 1115: J. Moulin Ollagnier, Ergodic Theory and Statistical Mechanics. VI, 147 pages. 1985.

Vol. 1116: S. Stolz, Hochzusammenhängende Mannigfaltigkeiten und ihre Ränder. XXIII, 134 Seiten. 1985.

Vol. 1117: D.J. Aldous, J.A. Ibragimov, J. Jacod, Ecole d'Été de Probabilités de Saint-Flour XIII – 1983. Édité par P.L. Hennequin. IX, 409 pages. 1985.

Vol. 1118: Grossissements de filtrations: exemples et applications. Seminaire, 1982/83. Edité par Th. Jeulin et M. Yor. V, 315 pages. 1985.

Vol. 1119: Recent Mathematical Methods in Dynamic Programming. Proceedings, 1984. Edited by I. Capuzzo Dolcetta, W.H. Fleming and T. Zolezzi. VI, 202 pages. 1985.

Vol. 1120: K. Jarosz, Perturbations of Banach Algebras. V, 118 pages. 1985.

Vol. 1121: Singularities and Constructive Methods for Their Treatment. Proceedings, 1983. Edited by P. Grisvard, W. Wendland and J.R. Whiteman. IX, 346 pages. 1985.

Vol. 1122: Number Theory. Proceedings, 1984. Edited by K. Alladi. VII, 217 pages. 1985.

Vol. 1123: Séminaire de Probabilités XIX 1983/84. Proceedings. Edité par J. Azéma et M. Yor. IV, 504 pages. 1985.

Vol. 1124: Algebraic Geometry, Sitges (Barcelona) 1983. Proceedings. Edited by E. Casas-Alvero, G.E. Welters and S. Xambó-Descamps. XI, 416 pages. 1985.

Vol. 1125: Dynamical Systems and Bifurcations. Proceedings, 1984. Edited by B.L.J. Braaksma, H.W. Broer and F. Takens. V, 129 pages. 1985.

Vol. 1126: Algebraic and Geometric Topology. Proceedings, 1983. Edited by A. Ranicki, N. Levitt and F. Quinn. V, 423 pages. 1985.

Vol. 1127: Numerical Methods in Fluid Dynamics. Seminar. Edited by F. Brezzi, VII, 333 pages. 1985.

Vol. 1128: J. Elschner, Singular Ordinary Differential Operators and Pseudodifferential Equations. 200 pages. 1985.

Vol. 1129: Numerical Analysis, Lancaster 1984. Proceedings. Edited by P.R. Turner. XIV, 179 pages. 1985.

Vol. 1130: Methods in Mathematical Logic. Proceedings, 1983. Edited by C.A. Di Prisco. VII, 407 pages. 1985.

Vol. 1131: K. Sundaresan, S. Swaminathan, Geometry and Nonlinear Analysis in Banach Spaces. III, 116 pages. 1985.

Vol. 1132: Operator Algebras and their Connections with Topology and Ergodic Theory. Proceedings, 1983. Edited by H. Araki, C.C. Moore, Ş. Strătilă and C. Voiculescu. VI, 594 pages. 1985.

Vol. 1133: K.C. Kiwiel, Methods of Descent for Nondifferentiable Optimization. VI, 362 pages. 1985.

Vol. 1134: G.P. Galdi, S. Rionero, Weighted Energy Methods in Fluid Dynamics and Elasticity. VII, 126 pages. 1985.

Vol. 1135: Number Theory, New York 1983–84. Seminar. Edited by D.V. Chudnovsky, G.V. Chudnovsky, H. Cohn and M.B. Nathanson. V, 283 pages. 1985.

Vol. 1136: Quantum Probability and Applications II. Proceedings, 1984. Edited by L. Accardi and W. von Waldenfels. VI, 534 pages. 1985.

Vol. 1137: Xiao G., Surfaces fibrées en courbes de genre deux. IX, 103 pages. 1985.

Vol. 1138: A. Ocneanu, Actions of Discrete Amenable Groups on von Neumann Algebras. V, 115 pages. 1985.

Vol. 1139: Differential Geometric Methods in Mathematical Physics. Proceedings, 1983. Edited by H. D. Doebner and J. D. Hennig. VI, 337 pages. 1985.

Vol. 1140: S. Donkin, Rational Representations of Algebraic Groups. VII, 254 pages. 1985.

Vol. 1141: Recursion Theory Week. Proceedings, 1984. Edited by H.-D. Ebbinghaus, G.H. Müller and G.E. Sacks. IX, 418 pages. 1985.

Vol. 1142: Orders and their Applications. Proceedings, 1984. Edited by I. Reiner and K. W. Roggenkamp. X, 306 pages. 1985.

Vol. 1143: A. Krieg, Modular Forms on Half-Spaces of Quaternions. XIII, 203 pages. 1985.

Vol. 1144: Knot Theory and Manifolds. Proceedings, 1983. Edited by D. Rolfsen. V, 163 pages. 1985.

Vol. 1145: G. W
1985.

Vol. 1146: Sémi
Proceedings, 1983–1984.
1985.

Vol. 1147: M. Wschebor, Surfaces Aléatoires. VII, 111 pages. 1985.

Vol. 1148: Mark A. Kon, Probability Distributions in Quantum Statistical Mechanics. V, 121 pages. 1985.

Vol. 1149: Universal Algebra and Lattice Theory. Proceedings, 1984. Edited by S. D. Comer. VI, 282 pages. 1985.

Vol. 1150: B. Kawohl, Rearrangements and Convexity of Level Sets in PDE. V, 136 pages. 1985.

Vol 1151: Ordinary and Partial Differential Equations. Proceedings, 1984. Edited by B.D. Sleeman and R.J. Jarvis. XIV, 357 pages. 1985.

Vol. 1152: H. Widom, Asymptotic Expansions for Pseudodifferential Operators on Bounded Domains. V, 150 pages. 1985.

Vol. 1153: Probability in Banach Spaces V. Proceedings, 1984. Edited by A. Beck, R. Dudley, M. Hahn, J. Kuelbs and M. Marcus. VI, 457 pages. 1985.

Vol. 1154: D.S. Naidu, A.K. Rao, Singular Pertubation Analysis of Discrete Control Systems. IX, 195 pages. 1985.

Vol. 1155: Stability Problems for Stochastic Models. Proceedings, 1984. Edited by V.V. Kalashnikov and V.M. Zolotarev. VI, 447 pages. 1985.

Vol. 1156: Global Differential Geometry and Global Analysis 1984. Proceedings, 1984. Edited by D. Ferus, R.B. Gardner, S. Helgason and U. Simon. V, 339 pages. 1985.

Vol. 1157: H. Levine, Classifying Immersions into $\mathbb{R}^4$ over Stable Maps of 3-Manifolds into $\mathbb{R}^2$. V, 163 pages. 1985.

Vol. 1158: Stochastic Processes – Mathematics and Physics. Proceedings, 1984. Edited by S. Albeverio, Ph. Blanchard and L. Streit. VI, 230 pages. 1986.

Vol. 1159: Schrödinger Operators, Como 1984. Seminar. Edited by S. Graffi. VIII, 272 pages. 1986.

Vol. 1160: J.-C. van der Meer, The Hamiltonian Hopf Bifurcation. VI, 115 pages. 1985.

Vol. 1161: Harmonic Mappings and Minimal Immersions, Montecatini 1984. Seminar. Edited by E. Giusti. VII, 285 pages. 1985.

Vol. 1162: S.J.L. van Eijndhoven, J. de Graaf, Trajectory Spaces, Generalized Functions and Unbounded Operators. IV, 272 pages. 1985.

Vol. 1163: Iteration Theory and its Functional Equations. Proceedings, 1984. Edited by R. Liedl, L. Reich and Gy. Targonski. VIII, 231 pages. 1985.

Vol. 1164: M. Meschiari, J.H. Rawnsley, S. Salamon, Geometry Seminar "Luigi Bianchi" II – 1984. Edited by E. Vesentini. VI, 224 pages. 1985.

Vol. 1165: Seminar on Deformations. Proceedings, 1982/84. Edited by J. Ławrynowicz. IX, 331 pages. 1985.

Vol. 1166: Banach Spaces. Proceedings, 1984. Edited by N. Kalton and E. Saab. VI, 199 pages. 1985.

Vol. 1167: Geometry and Topology. Proceedings, 1983–84. Edited by J. Alexander and J. Harer. VI, 292 pages. 1985.

Vol. 1168: S.S. Agaian, Hadamard Matrices and their Applications. III, 227 pages. 1985.

Vol. 1169: W.A. Light, E.W. Cheney, Approximation Theory in Tensor Product Spaces. VII, 157 pages. 1985.

Vol. 1170: B.S. Thomson, Real Functions. VII, 229 pages. 1985.

Vol. 1171: Polynômes Orthogonaux et Applications. Proceedings, 1984. Edité par C. Brezinski, A. Draux, A.P. Magnus, P. Maroni et A. Ronveaux. XXXVII, 584 pages. 1985.

Vol. 1172: Algebraic Topology, Göttingen 1984. Proceedings. Edited by L. Smith. VI, 209 pages. 1985.

M. Knebusch, Locally Semialgebraic Spaces. XVI,

es in Continuum Physics, Buffalo 1982. Seminar. vere and S.H. Schanuel. V, 126 pages. 1986.

K. Mathiak, Valuations of Skew Fields and Projective Hjel Spaces. VII, 116pages. 1986.

Vol. 1176: R.R. Bruner, J.P. May, J.E. McClure, M. Steinberger, H_∞ Ring Spectra and their Applications. VII, 388 pages. 1986.

Vol. 1177: Representation Theory I. Finite Dimensional Algebras. Proceedings, 1984. Edited by V. Dlab, P. Gabriel and G. Michler. XV, 340 pages. 1986.

Vol. 1178: Representation Theory II. Groups and Orders. Proceedings, 1984. Edited by V. Dlab, P. Gabriel and G. Michler. XV, 370 pages. 1986.

Vol. 1179: Shi J.-Y. The Kazhdan-Lusztig Cells in Certain Affine Weyl Groups. X, 307 pages. 1986.

Vol. 1180: R. Carmona, H. Kesten, J.B. Walsh, École d'Été de Probabilités de Saint-Flour XIV – 1984. Édité par P.L. Hennequin. X, 438 pages. 1986.

Vol. 1181: Buildings and the Geometry of Diagrams, Como 1984. Seminar. Edited by L. Rosati. VII, 277 pages. 1986.

Vol. 1182: S. Shelah, Around Classification Theory of Models. VII, 279 pages. 1986.

Vol. 1183: Algebra, Algebraic Topology and their Interactions. Proceedings, 1983. Edited by J.-E. Roos. XI, 396 pages. 1986.

Vol. 1184: W. Arendt, A. Grabosch, G. Greiner, U. Groh, H.P. Lotz, U. Moustakas, R. Nagel, F. Neubrander, U. Schlotterbeck, One-parameter Semigroups of Positive Operators. Edited by R. Nagel. X, 460 pages. 1986.

Vol. 1185: Group Theory, Beijing 1984. Proceedings. Edited by Tuan H.F. V, 403 pages. 1986.

Vol. 1186: Lyapunov Exponents. Proceedings, 1984. Edited by L. Arnold and V. Wihstutz. VI, 374 pages. 1986.

Vol. 1187: Y. Diers, Categories of Boolean Sheaves of Simple Algebras. VI, 168 pages. 1986.

Vol. 1188: Fonctions de Plusieurs Variables Complexes V. Séminaire, 1979–85. Edité par François Norguet. VI, 306 pages. 1986.

Vol. 1189: J. Lukeš, J. Malý, L. Zajíček, Fine Topology Methods in Real Analysis and Potential Theory. X, 472 pages. 1986.

Vol. 1190: Optimization and Related Fields. Proceedings, 1984. Edited by R. Conti, E. De Giorgi and F. Giannessi. VIII, 419 pages. 1986.

Vol. 1191: A.R. Its, V.Yu. Novokshenov, The Isomonodromic Deformation Method in the Theory of Painlevé Equations. IV, 313 pages. 1986.

Vol. 1192: Equadiff 6. Proceedings, 1985. Edited by J. Vosmansky and M. Zlámal. XXIII, 404 pages. 1986.

Vol. 1193: Geometrical and Statistical Aspects of Probability in Banach Spaces. Proceedings, 1985. Edited by X. Femique, B. Heinkel, M.B. Marcus and P.A. Meyer. IV, 128 pages. 1986.

Vol. 1194: Complex Analysis and Algebraic Geometry. Proceedings, 1985. Edited by H. Grauert. VI, 235 pages. 1986.

Vol. 1195: J.M. Barbosa, A.G. Colares, Minimal Surfaces in $\mathbb{R}^3$. X, 124 pages. 1986.

Vol. 1196: E. Casas-Alvero, S. Xambó-Descamps, The Enumerative Theory of Conics after Halphen. IX, 130 pages. 1986.

Vol. 1197: Ring Theory. Proceedings, 1985. Edited by F.M.J. van Oystaeyen. V, 231 pages. 1986.

Vol. 1198: Séminaire d'Analyse, P. Lelong – P. Dolbeault – H. Skoda. Seminar 1983/84. X, 260 pages. 1986.

Vol. 1199: Analytic Theory of Continued Fractions II. Proceedings, 1985. Edited by W.J. Thron. VI, 299 pages. 1986.

Vol. 1200: V.D. Milman, G. Schechtman, Asymptotic Theory of Finite Dimensional Normed Spaces. With an Appendix by M. Gromov. VIII, 156 pages. 1986.